A.P.2210G—P.N.

PILOT'S NOTES

METEOR Mk. 7

Prepared by Direction
of the
Minister of Supply

Promulgated by Command
of the
Air Council

4th Edition
December, 1957

A.P.2210G—P.N.

PILOT'S NOTES

METEOR Mk. 7

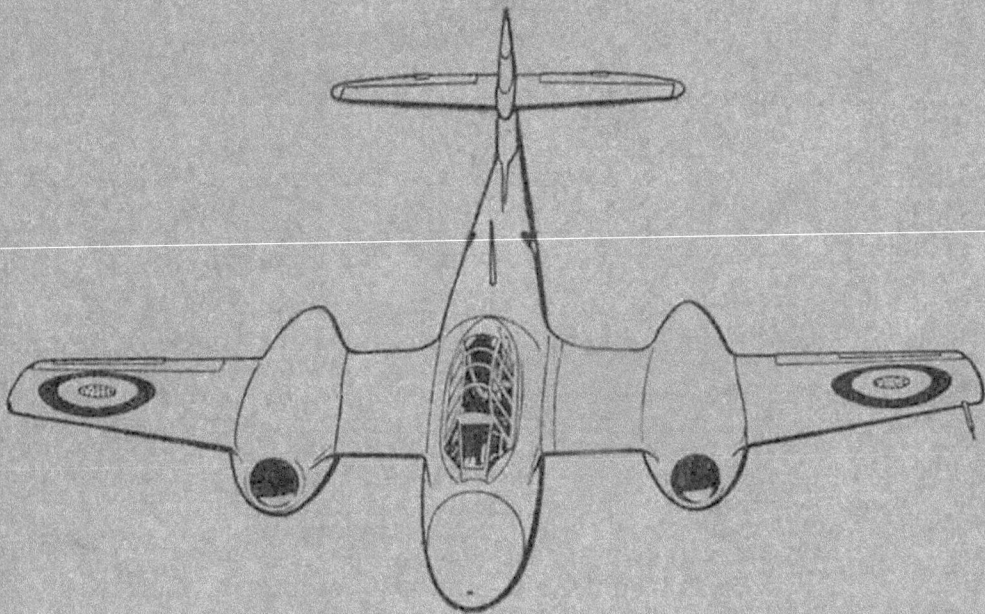

RESTRICTED

MINISTRY OF DEFENCE
August. 1970

Amendment List No. 6
to A.P.2210G—P.N.
Pilot's Notes

METEOR Mk. 7

NOTE 1: — This Amendment List includes information covering the following Special Flying Instruction and Modifications:

(a) SFI/RN/11/70

(b) Mods 976, 1910, 011/FC, 05/FEAF

NOTE 2:— When a manuscript amendment is made, *endorse* the adjacent margin "AL6".

NOTE 3:— When the Amendment List is fully incorporated:—

(a) *Affix* this sheet to the inside front cover of the Notes.

(b) *Certify* its incorporation on Page 1 of the Notes.

PAGE	PARA.	AMENDMENT
5	LIST OF CONTENTS	Below "ASYMMETRIC FLYING" *insert:* "Asymmetric flying, general......73A".
14	11 (a)	*Add* at end: "Mod 976 repositions the inertia switch in the fuselage".
17	16 (a)	*Add* at end: ", or between the fuel gauges post-Mod 05/FEAF".
19	19	*Add* at end: "Mod 011/FC fits a handle to the rudder trim wheel".
22	25 (*contd.*)	After sub-para. (h) *insert:* "(j) Mod 1910 introduces a fatigue meter Mk. 14".
23	27 (b) (iv) lines, 3, 4	*Delete* from "when" to end of sentence and *substitute* "If the rear hooks fail to disengage".
24	27 (*contd.*), 28	*Amend* by gummed slip.
30	46 (a)	*Amend* by gummed slip.
47	—	*Amend* by gummed page.
49	65 Line 7	*Delete* "Stall turn......280–300".
53	—	*Amend* by gummed page.
54	75 (b)	*Add* at end "But see Para. 73A (c)".
61	—	*Amend* by gummed page.

RESTRICTED

(1287) Dd. 728103 1,162 8/70 H.P. Ltd. 1620/1

NOTES TO USERS

These Notes are complementary to A.P.129 (6th Edition), Flying, and it is assumed that all concerned have a thorough knowledge of the chapters which are relevant to the operation of this type of aircraft.

Additional copies may be obtained by the station publications officer, by application on R.A.F. Form 294A, in quadruplicate, to Command Headquarters for onward transmission to A.P.F.S. (see A.P.113A). The number of this publication must be quoted in full—A.P.2210G—P.N. (4th Edition).

Comments and suggestions should be forwarded to the Officer Commanding, Handling Squadron, Royal Air Force, Boscombe Down, Wilts.

AMENDMENTS

Amendment lists will be issued as necessary and will be gummed for affixing to the inside front cover of these Notes.

Each amendment list will, where applicable, be accompanied by gummed slips for sticking in the appropriate places in the text. Incorporation of an amendment list must be certified by inserting date of incorporation and initials below.

A.L. No.	INITIALS	DATE	A.L. No.	INITIALS	DATE
1			4		
2			5		
3			6		

METEOR Mk. 7

LIST OF ASSOCIATED PUBLICATIONS

Title	A.P. No.
Meteor 7, general and technical information..	2210G. Vol. 1
Derwent 5, 8 and 9 descriptive handbook ..	4038B, C & D. Vol. 1
Fuel system components for gas-turbine aero-engines ..	4282 series
Rotol accessory gear boxes and drives	2240A
Electrical equipment manual ..	1095 series
Aircraft hydraulic equipment ..	1803 series
Aircraft wheels, tyres and brakes ..	2337
R.A.F. signal manual ..	1186 series
Instrument manual ..	1275 series
Safety equipment ..	1182 series
Cine cameras and accessories ..	1355D
Fire prevention and fire extinguishing equipment in aircraft ..	957C
Air pump units ..	1519

1 C 45B CAMERA	8 RADIO	15 LANDING LAMP
2 BALLAST WEIGHTS	9 FIRE EXTINGUISHER BOTTLES	16 DERWENT TURBINE
3 RETRACTABLE GYRO GUNSIGHT	10 AIR CONTAINER	17 OIL TANK
4 PUPIL PILOT	11 DISTANT-READING COMPASS	18 VACUUM PUMP
5 HINGED HOOD	12 EMERGENCY LANDING SKID	19 ACCESSORY GEARBOX
6 INSTRUCTOR PILOT	13 AIR BRAKE FLAPS	20 HYMATIC COMPRESSOR
7 FUEL TANK	14 JET PIPE	21 STARTER MOTOR

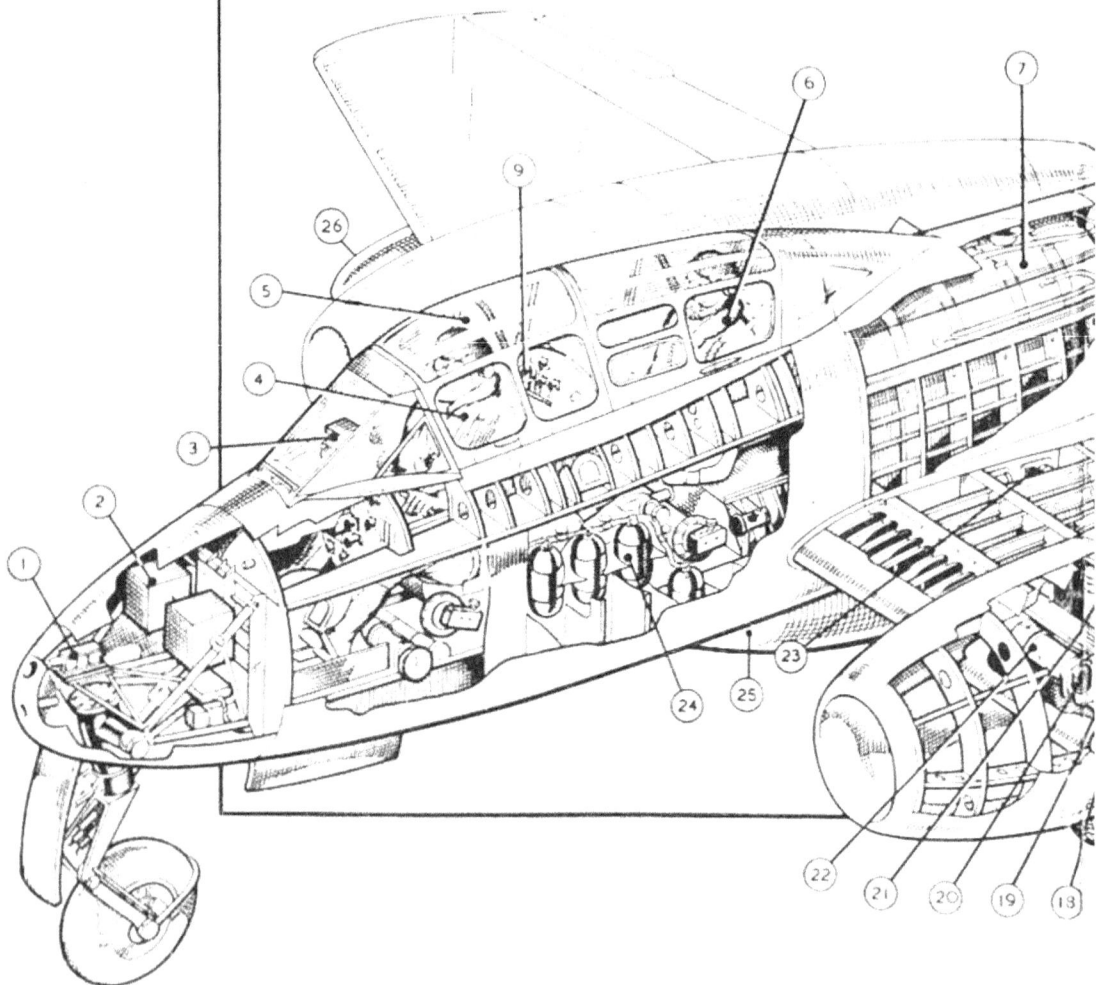

22 GENERATOR

23 GROUND/FLIGHT SWITCH & GROUND SOCKET

24 OXYGEN BOTTLES

25 VENTRAL DROP TANK

26 WING DROP TANK

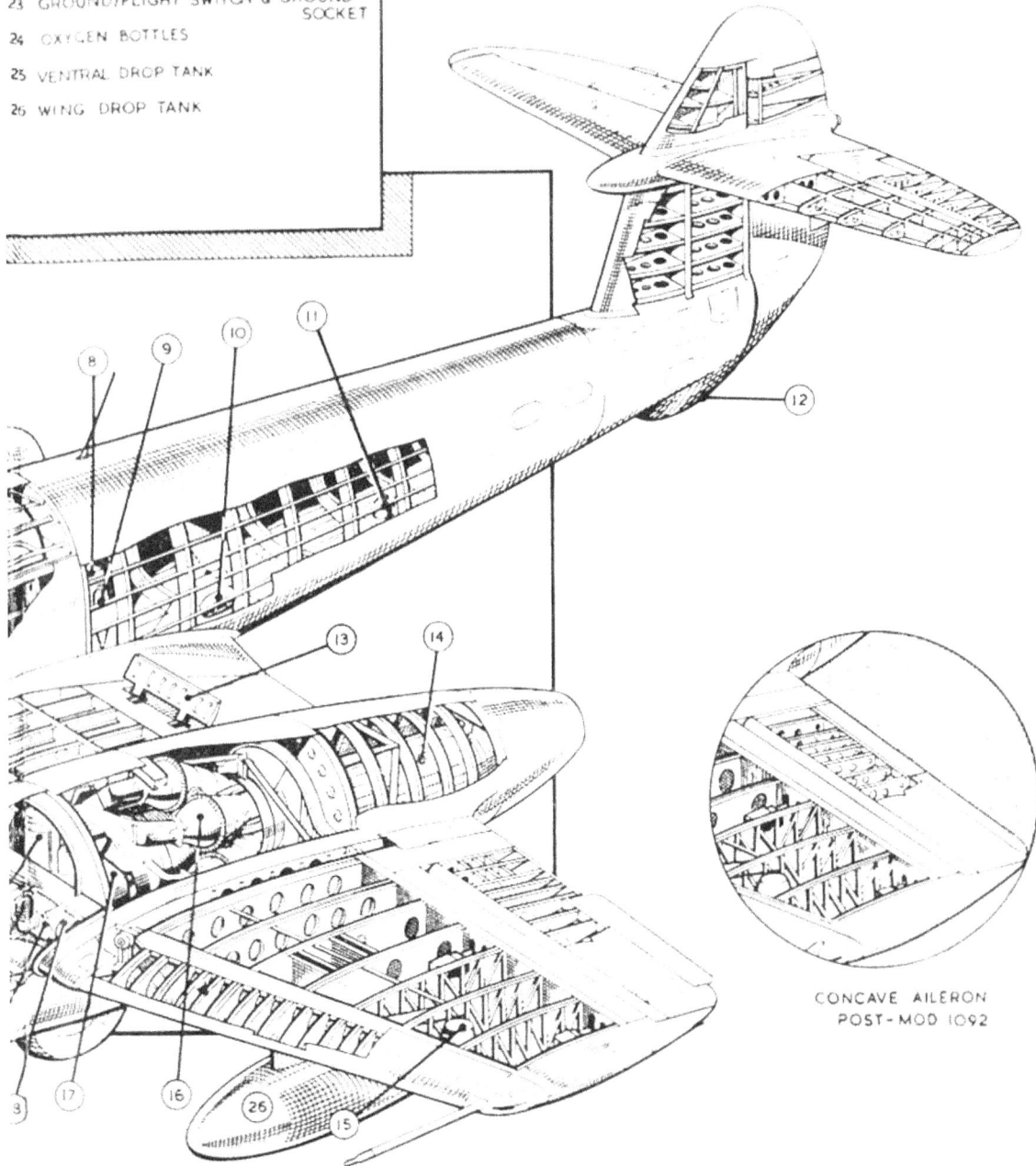

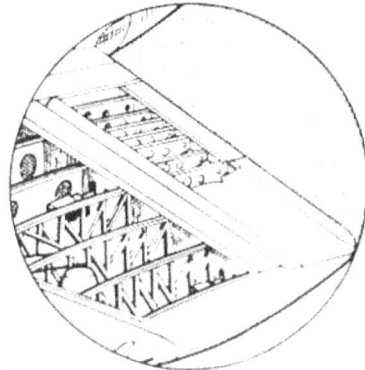

CONCAVE AILERON
POST-MOD 1092

METEOR T. Mk. 7

Air Ministry
December, 1957

A.P.2210G—P.N.
Pilot's Notes
4th Edition

METEOR MK. 7

NOTE. This edition cancels the 3rd edition, issued in December 1954.

LIST OF CONTENTS

PART I—DESCRIPTIVE

INTRODUCTION

PART V—OPERATING DATA

PART VI—ILLUSTRATIONS

PART I

DESCRIPTIVE

NOTE.—Throughout this publication the following conventions apply:—

(a) Words in capital letters indicate the actual markings on the controls concerned.

(b) The numbers quoted in brackets after items in the text refer to the illustrations in Part VI.

(c) Unless otherwise stated all airspeeds and mach numbers quoted are indicated.

INTRODUCTION

(a) The Meteor Mk. 7 is a two-seat, jet-propelled, trainer aircraft, powered by two Derwent Mk. 8 or Mk. 9 engines. In the latter engines, the priming pumps and torch-igniters are replaced by high energy igniter units, which permit relighting at higher altitudes and greater forward speeds. The engines may have either large or small intakes fitted.

(b) The aircraft is unarmed but may be fitted with a gyro-gunsight and camera. The cockpit is unpressurised. Long-range fuel tanks can be fitted, giving a maximum range at altitude of approximately 780 nautical miles.

(c) The principal aircraft dimensions are as follows:—

Wing span	37 ft 2 ins.
Length	43 ft. 6 ins.
Maximum height above ground	13 ft. 0 ins.
Wheel track	10 ft. 5 ins.

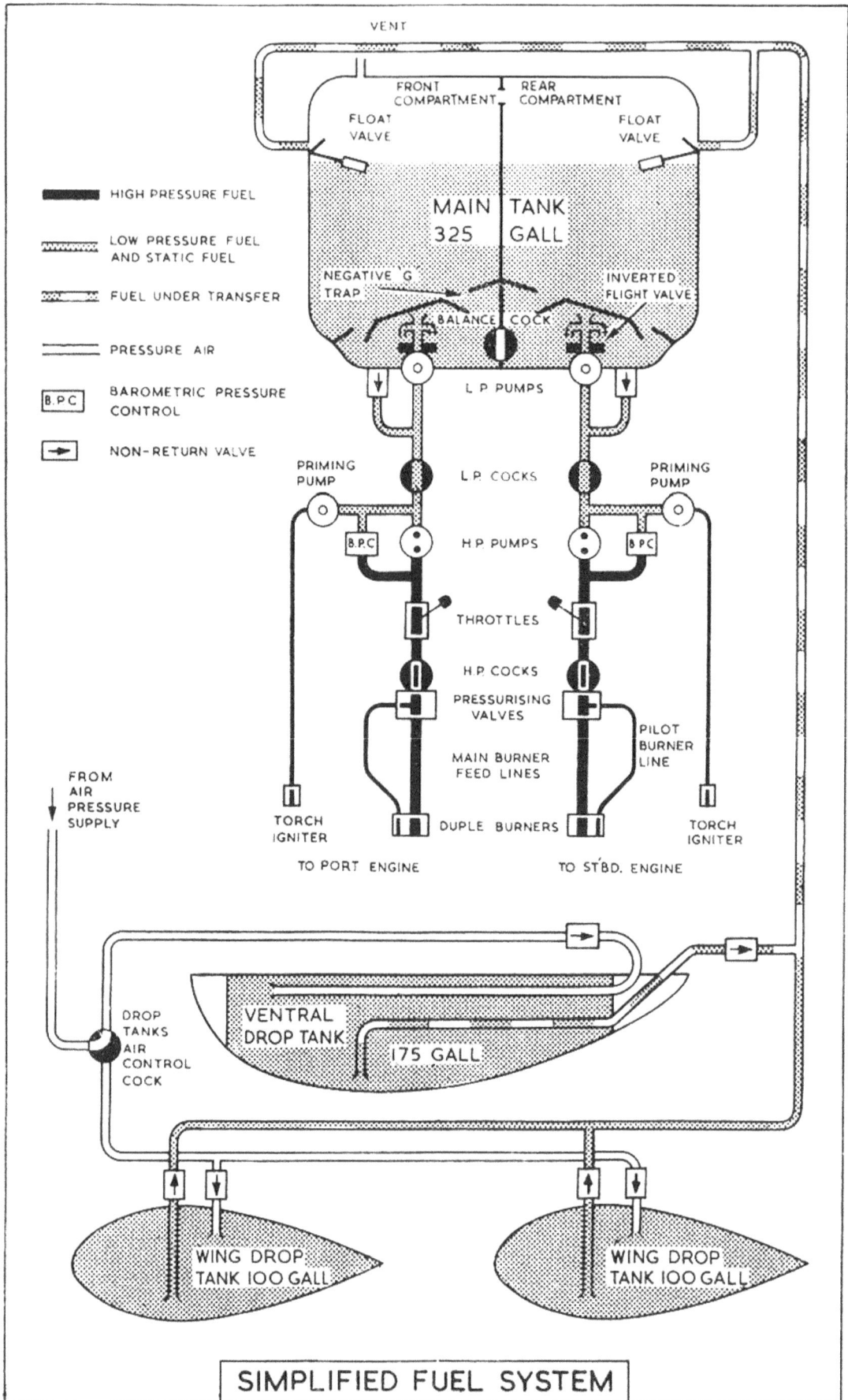

VENT

FRONT COMPARTMENT | REAR COMPARTMENT

FLOAT VALVE   FLOAT VALVE

HIGH PRESSURE FUEL

LOW PRESSURE FUEL AND STATIC FUEL

FUEL UNDER TRANSFER

PRESSURE AIR

B.P.C   BAROMETRIC PRESSURE CONTROL

→   NON-RETURN VALVE

MAIN TANK 325 GALL

NEGATIVE 'G' TRAP

INVERTED FLIGHT VALVE

BALANCE COCK

L P PUMPS

PRIMING PUMP   PRIMING PUMP

L.P. COCKS

B.P.C   B PC

H.P. PUMPS

THROTTLES

H.P. COCKS

PRESSURISING VALVES

FROM AIR PRESSURE SUPPLY

MAIN BURNER FEED LINES

PILOT BURNER LINE

TORCH IGNITER

DUPLE BURNERS

TORCH IGNITER

TO PORT ENGINE   TO ST'BD. ENGINE

DROP TANKS AIR CONTROL COCK

VENTRAL DROP TANK

175 GALL

WING DROP TANK 100 GALL

WING DROP TANK 100 GALL

SIMPLIFIED FUEL SYSTEM

Derwent Mk. 8 engines

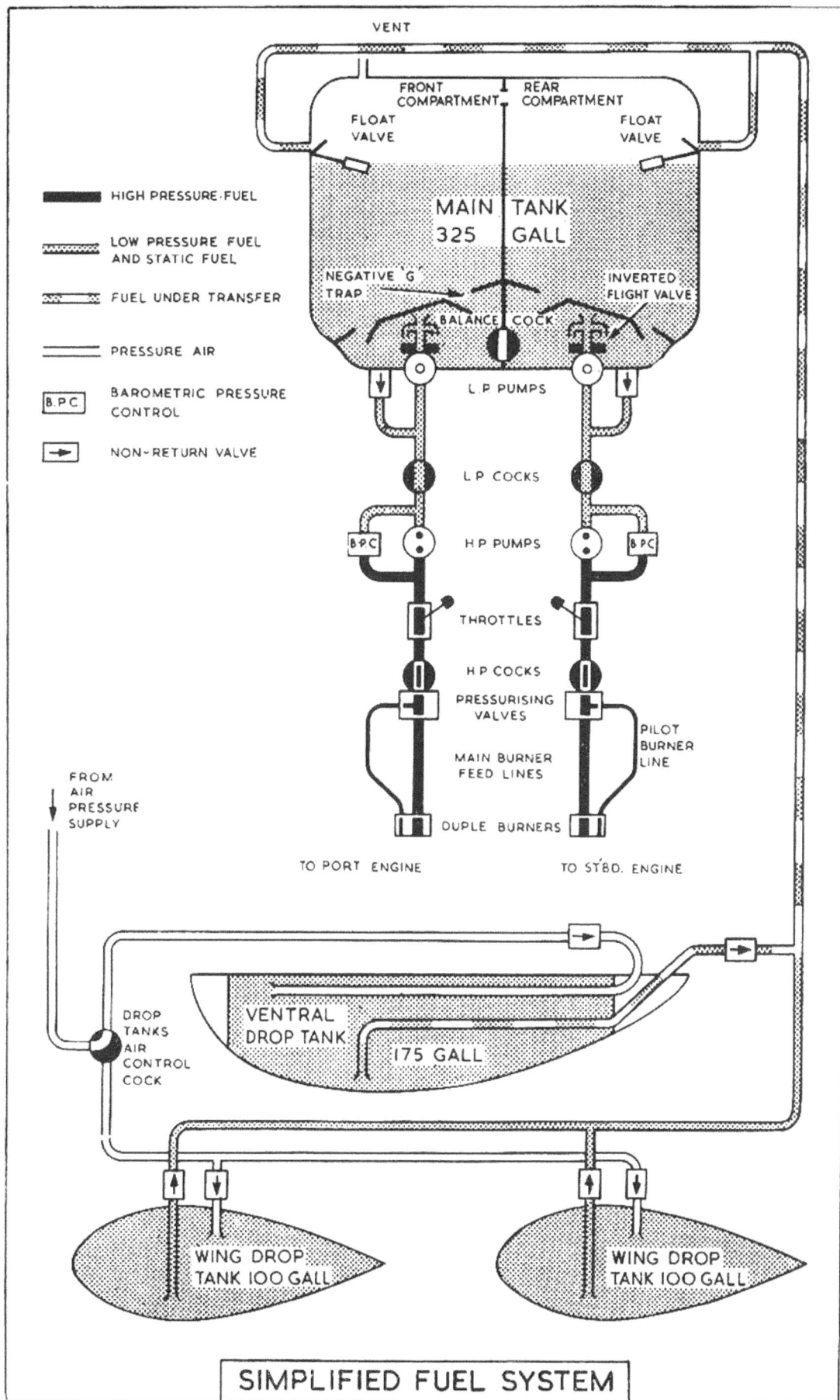

VENT

FRONT COMPARTMENT   REAR COMPARTMENT

FLOAT VALVE   FLOAT VALVE

MAIN TANK 325 GALL

NEGATIVE 'G' TRAP   INVERTED FLIGHT VALVE

BALANCE COCK

L.P. PUMPS

HIGH PRESSURE FUEL

LOW PRESSURE FUEL AND STATIC FUEL

FUEL UNDER TRANSFER

PRESSURE AIR

B.P.C.   BAROMETRIC PRESSURE CONTROL

NON-RETURN VALVE

L.P. COCKS

B.P.C.   H.P. PUMPS   B.P.C.

THROTTLES

H.P. COCKS

PRESSURISING VALVES

MAIN BURNER FEED LINES

PILOT BURNER LINE

FROM AIR PRESSURE SUPPLY

DUPLE BURNERS

TO PORT ENGINE   TO ST'BD. ENGINE

DROP TANKS AIR CONTROL COCK

VENTRAL DROP TANK   175 GALL

WING DROP TANK 100 GALL   **WING DROP TANK 100 GALL**

SIMPLIFIED FUEL SYSTEM

Derwent Mk. 9 engines

FUEL AND OIL SYSTEMS

1. **Fuel tanks and gauges**

(a) *Fuselage tank*

The fuselage fuel tank is divided into two equal compartments. The front compartment normally feeds the port engine and the rear compartment the starboard engine but the compartments can be interconnected by the operation of a balance cock (29). Each compartment contains negative G traps, ensuring sufficient fuel for 15 seconds' inverted flight. There are vent holes near the top of the dividing diaphragm and fuel may, in certain circumstances pass from one compartment to another through these holes.

(b) *Drop tanks*

A ventral tank and two under-wing drop tanks may be carried.

(c) *Tank capacities*

Tank capacities (in gallons) are as follows:—

Fuselage tank	325
Ventral tank	175
Wing drop tanks (2 × 100)	200
Total fuel	700

(d) *Fuel gauges*

Two electrical fuel contents gauges (52, 54), one for each compartment of the fuselage tank, are below the front cockpit instrument panel and similar gauges (111) are below the rear cockpit instrument panel. There are no gauges for the drop tanks.

2. **Fuel transfer**

Fuel from the drop tanks is transferred to the fuselage tank by air pressure from the engines. Air pressure to either the ventral tank or the wing tanks is selected by a T-handle (35) at the upper left-hand side of the front cockpit instrument panel. (This handle is also used to jettison the ventral tank.) A float valve in each fuselage tank compartment prevents overfilling of the fuselage tank by fuel from the drop tanks: the fuel level in the fuselage tank falls to approximately 125 gallons a side before the float valves open.

3. **Main fuel feed**

(a) *Low pressure*

The L.P. pumps at the base of each compartment of the main tank deliver fuel via the L.P. cocks to the H.P. pumps. Should an L.P. pump fail, fuel will by-pass the pump and be fed to the H.P. pump by gravity, via non-return valves.

(b) *High pressure*

From the H.P. pump, fuel flows at high pressure to the throttle valve and the barometric pressure control (B.P.C.). The B.P.C. maintains the correct fuel-air ratio for all altitudes, speeds and throttle settings. From the throttle valve, fuel passes through the H.P. cocks and then through pressurising valves to the main and pilot burner lines and thence to the duple burners. When pressure is low (i.e. at low r.p.m.), the pressurising valve feeds fuel to the burners through the pilot line only and, when pressure rises, through both the main and pilot lines.

(c) *Starting and relighting fuel supply (Derwent Mk. 8 only)*

On the Derwent Mk. 8 engine, fuel for starting and relighting is drawn from the main line between the L.P. cock and the H.P. pump and passes to the torch igniter via a priming pump, which is automatically operated when either the starting or the relight button is pressed.

4. **Fuel controls**

(a) *L.P. pumps circuit breakers*

The circuit breakers (19) for the port and starboard pumps are at the forward end of the port shelf in the front cockpit. These circuit breakers must be made to complete the engine-starting circuit and should remain made at all times when the engine is running. Test pushbuttons (24) and an ammeter socket (21) are provided for testing the pump output.

(b) *L.P. cock levers*

The L.P. cock levers are at the rear of each cockpit, one on either side of the seat. In the front cockpit the levers (1, 80) are the inner ones of the pair on each side of the seat; in the rear cockpit the levers (87, 124) are the rear ones of each pair. In the off (up) position the L.P. cocks cut off the flow of fuel from the main tank and should not be used to stop the engines, except in an emergency, as their use may damage the H.P. pumps and aerate the fuel system.

(c) *H.P. cock lever*

The H.P. cock levers are on either side of the seat in each cockpit, beside the L.P. cocks ((3, 79) in the front cockpit, (89, 123) in the rear cockpit). In the OFF (up) position, they cut off the supply of fuel to the burners and should always be used for stopping the engines. The appropriate one must also be closed in the event of engine failure.

(d) *Balance cock*

The balance cock (29) is on the floor of the front cockpit, just aft of the rudder trimming tab control. It is pulled up to interconnect the two compartments of the main tank.

(e) *Drop tank controls*

The transfer of fuel from the drop tanks is controlled by a T-handle (35) at the top left-hand side of the front cockpit instrument panel. The handle is marked PULL TO JETTISON and is inscribed with an arrow. Pre-Mod. 1841, the handle has three positions, WING ON, BELLY ON and OFF. When set to WING ON, fuel is transferred from the wing drop tanks and, when set to BELLY ON, from the ventral tank. Mod. 1841 deletes the OFF position. A red warning light (30), at the bottom left of the front cockpit instrument panel, comes on whenever the pressure in the transfer lines falls below 1 lb./sq. in. Pre-Mod. 1841 this light goes out when the handle is moved to OFF.

5. **Drop tanks jettison controls**

(a) *Ventral tank*

The ventral tank is jettisoned by pulling out the T-handle (35) from any position. The handle is marked PULL TO JETTISON.

(b) *Wing drop tanks*

The wing drop tanks are jettisoned by pulling back the lever (83) on the starboard side of the front cockpit.

6. **Oil system**

An oil tank is fitted in each engine nacelle. Pre-Derwent Mod. 571, each tank has a capacity of 22 pints of oil with a 7-pint air space. Derwent Mod. 571 reduces the oil capacity to $19\frac{1}{2}$ pints.

ENGINE CONTROLS

7. **Derwent Mk. 8 and 9 engines**

(a) Both the Derwent Mk. 8 and 9 engines are centrifugal-flow jet engines giving approximately 3,500 lb. static thrust at sea level under I.S.A. conditions. The Mk. 8 engine has priming pumps and torch igniters, while in the Mk. 9 engine these are deleted and high energy igniter units are introduced.

(b) The maximum engine r.p.m. are governed at 14,550 at sea level; the r.p.m. will increase with altitude to 14,700 r.p.m.

8. **Throttle controls**

Throttle levers (12) and (90), are in slides on the port wall of both cockpits and, when fully shut (back), sufficient fuel is allowed to pass to enable the engines to idle. No friction damping control is fitted. The throttles must be fully closed before the engine starting circuit is complete. In the front cockpit, the port throttle grip incorporates the G.G.S. ranging control. The inboard throttle may carry the press-to-transmit switch and a spring-loaded mute switch.

9. **Engine starting controls**

(a) The starting cycle is automatically controlled by time switches. When the L.P. pump is on and the throttle closed (to operate the micro-switch (7)), pressing and releasing the shielded starter pushbuttons (20) on the port shelf completes the starting circuit and actuates the time switches, which in turn bring the starter motors and igniters into circuit. Current to the starter circuit is automatically cut off after 30 seconds.

(b) *Derwent Mk. 8 engines.*

An ignition isolating switch and a priming pump isolating switch for each engine are fitted in the appropriate under-carriage bay. With the priming pump isolating switch OFF, the pump is isolated. With the ignition isolating switch OFF, the priming pump, the torch igniters and the booster coils are all isolated. This allows the engine to be blown through after a false start, using the normal starting gear.

(c) *Derwent Mk. 9 engines*

Two switches, labelled MAIN SUPPLY and HIGH ENERGY IGNITER SUPPLY, are fitted in each under-carriage bay. Either switch isolates the high energy igniter unit from the electrical supply; in addition, the main supply switch isolates the igniter line relay. With either switch off, the engine may be blown through after a false start, using the normal starting gear.

10. Relighting controls

A relighting pushbutton (4, 78, 88, 122) is in the top of each H.P. cock lever. In some aircraft the pushbuttons may be on a panel on the port side of each cockpit. The push-buttons are used for relighting the engines during flight or for ground testing the priming pumps and torch igniters (Derwent Mk. 8) or high energy ignition units (Derwent Mk. 9). The relight buttons must always be used for relighting in flight and never the normal starting system.

11. Engine fire extinguishers

(a) There are two engine fire extinguisher bottles in the rear fuselage. Each bottle operates through a spray ring on the appropriate engine, whenever the associated fire extin-guisher pushbutton (45) in the front cockpit is pressed. In the event of a crash landing, an inertia switch in the starboard nacelle operates both bottles.

(b) Engine fire-warning lights (36, 43), one for each engine, are on the port and starboard glare shields in the front cockpit; the lights are operated by fusible flame switches in the engine nacelles; the lights will not go out when the fire is extinguished.

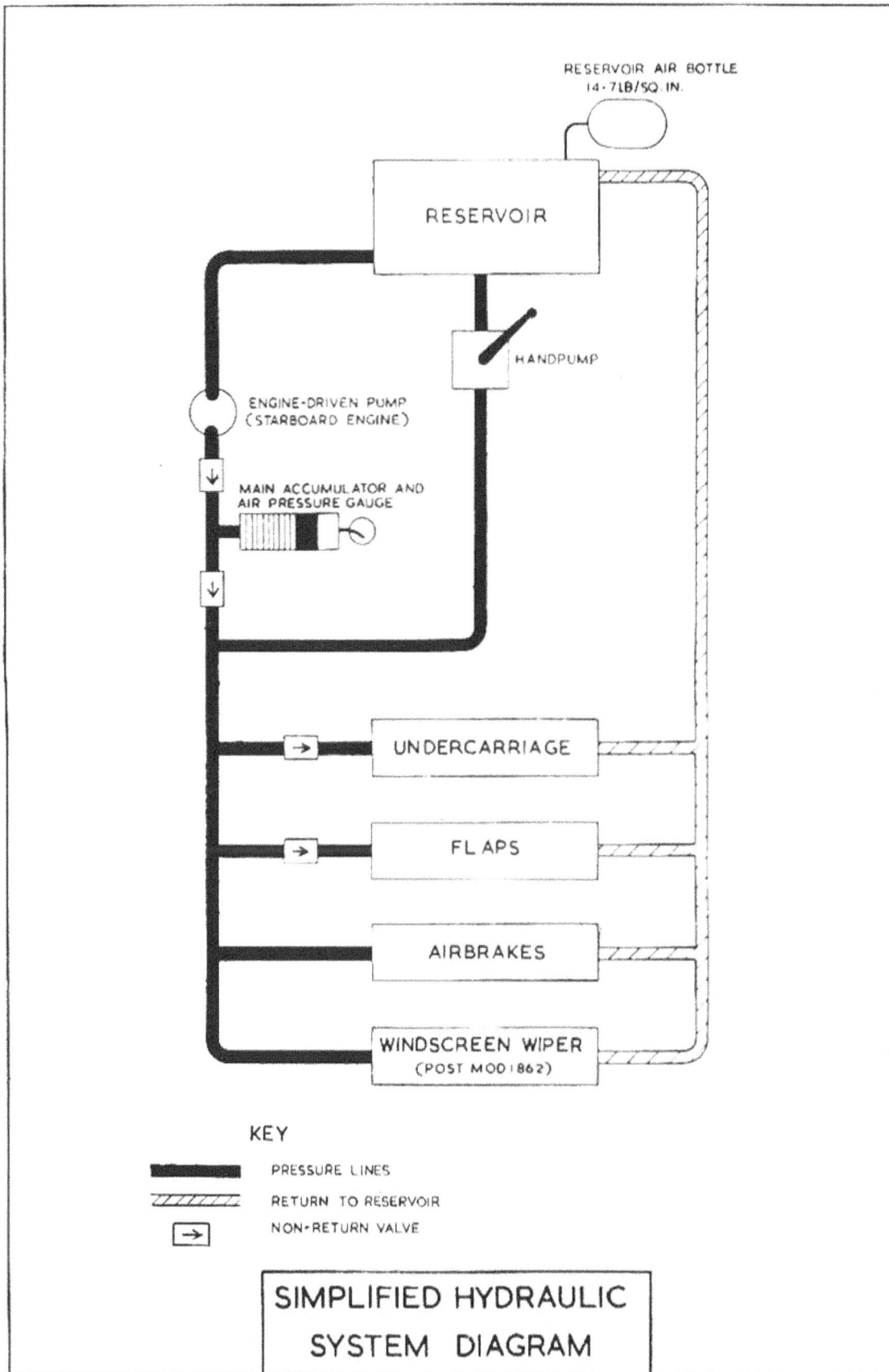

RESERVOIR AIR BOTTLE
14·7LB/SQ.IN.

RESERVOIR

HANDPUMP

ENGINE-DRIVEN PUMP
(STARBOARD ENGINE)

MAIN ACCUMULATOR AND
AIR PRESSURE GAUGE

UNDERCARRIAGE

FLAPS

AIRBRAKES

WINDSCREEN WIPER
(POST MOD1862)

KEY

PRESSURE LINES

RETURN TO RESERVOIR

→ NON-RETURN VALVE

SIMPLIFIED HYDRAULIC
SYSTEM DIAGRAM

12. **Engine instruments**

The following engine instruments are provided:—

R.p.m. indicators	Both cockpits
Jet pipe temperature gauges	Front cockpit in all aircraft. In rear cockpit (94) on post-Mod. 1656 aircraft.
Oil pressure gauges (51, 55)	Front cockpit only.

MAIN SERVICES

13. **Hydraulic system**

(a) A hydraulic pump, driven by the starboard engine, provides pressure for the operation of the following:—

Undercarriage
Flaps
Airbrakes
Windscreen wiper (post Mod. 1862)

Page 16
Para.
13(b)
A.L.3.

(b) The engine-driven pump also charges an accumulator, which provides a reserve of pressure for operating any of the services when the engine is not running. The accumulator air inflation point and pressure gauge are on the rear cockpit floor, on the starboard side of the seat. The correct air pressure is 800 lb./sq. in.

NOTE.—In tropical temperatures, the efficiency of the hydraulic accumulator may be reduced.

(c) An air bottle, connected to the reservoir, maintains sea level atmospheric pressure in the tank at any altitude.

(d) A handpump (85), to starboard of the front cockpit seat, will operate all the services through the normal pipe lines, after normal selection. The hand pump cannot be used to charge the accumulator.

(e) There is no hydraulic pressure gauge, other than the accumulator gauge in the rear cockpit.

14. **Pneumatic system**

(a) A compressor on the port engine charges an air storage cylinder in the rear fuselage to a pressure of 450 lb./sq. in. for the operation of the brakes. A pressure reducing valve in the system reduces the operating pressure to 220 lb./sq. in. This pressure is further reduced to 120 lb./sq. in. by the brakes relay.

(b) A triple-pressure gauge (8) on the front cockpit port wall shows the pressure in the air storage cylinder and the pressure available at each brake.

15. Vacuum system

A vacuum pump on each engine provides suction for the operation of the artificial horizons, direction indicators (rear cockpit only, post Mod. 1177) and turn and slip indicators (pre-Mod. 1228). If one pump is not running, the other pump output is sufficient to serve all instruments; there is no changeover cock or suction gauge.

16. Electrical system

(a) *D.C. supply*

Two 3 kw generators, one on each engine, supply the 24-volt electrical system and charge the two 12-volt aircraft batteries which are connected in series. The generator cut-in speed is approximately 5,000 r.p.m. Two generator failure warning lights may be on the front cockpit instrument panel (post Mod. 1225)

(b) *External battery*

The external battery socket is in the port wheel well, with the ground/flight switch beside it. Also at this station is a voltmeter (or, post Mod. 1431, a voltmeter socket) to measure the output of the generator, aircraft batteries or external battery. The ground/flight switch, when set to GROUND, isolates the batteries and the generators from all electrical services except, post Mod. 1537, the high energy ignition circuit (Derwent Mk. 9). On Post-Mod. 1537 aircraft, the peak starting load is taken by the internal batteries while, with the ground/flight switch at GROUND, the remainder of the load is taken by the external battery.

NoTE.—If the external battery is disconnected before the ground/flight switch is moved to FLIGHT, the j.p.t. gauge may be damaged.

(c) *Mk. 4F compass supplies* (post-Mod. 1177)

    (i) Alternating current for the Mk. 4F compass is supplied by one of two inverters in the rear fuselage, inverter A being the main inverter and inverter B the standby.

    (ii) D.C. for the compass and inverters is controlled by a master switch (59) and a D.C. NORMAL/EMERGENCY switch (65), on the starboard switch panel in the front cockpit. When the latter switch is moved to EMERGENCY, it connects an alternative D.C. supply to the inverters and renders the master switch inoperative.

    (iii) The A.C. supply from the inverters is controlled by the A.C. NORMAL/EMERGENCY switch (64) beside the D.C. switch. An A.C. voltmeter (53), below the compass, registers in the white sector of the dial when A.C. is being supplied to the compass. If the voltmeter is not registering in the white sector, inverter failure is indicated and the A.C. switch should be put to EMERGENCY to bring in inverter B.

AIRCRAFT CONTROLS

17. Flying controls

(a) The flying controls are conventional: the rudder pedals in each cockpit can be adjusted for reach after pulling out the knob (33) (104) on the left-hand side of each panel. The pedals are mechanically interconnected to ensure equal adjustment.

(b) On aircraft embodying Mod. 1092, the ailerons are fitted with spring tabs in addition to geared tabs.

18. Flying controls locking gear and picketing rings

(a) The flying controls are locked in the neutral position by four rods which have small pegs at each end. These rods are fitted on the controls in the front cockpit, two of them being used to lock the rudder pedals to the elevator torque tube and the other two to connect the control column to attachment points (86) on the cockpit starboard wall and the bulkhead behind the pilot's seat. When not in use, the locking gear is stowed on the decking between the two cockpits.

(b) Removable picketing rings, stowed on the decking between the two cockpits, screw into holes in the underside of the mainplanes and the emergency skid; the holes are normally plugged with screws.

19. Trimming controls

The elevator trimming tabs are controlled by handwheels, (25) (97), on the port side of both cockpits. The rudder trimming tab is controlled by smaller handwheels, (28) (100), aft and to the left of each elevator trimming tab handwheel. Both elevator and rudder trimming tab controls work in the natural sense and each has an adjacent indicator (26), (27), (98) and (99).

20. Undercarriage controls

(a) *Normal operation*

The hydraulically-operated undercarriage is selected by the levers (31) (102), one in each cockpit. The levers have two positions, UP and DOWN.

(b) *Ground lock*

When the weight of the aircraft is on the wheels or when the port oleo is compressed, the selector lever is locked down and it is not possible to select UP. Mod. 1335 introduces an override lever (15) (91) on the port wall of each cockpit. The lever is marked EMERGENCY ONLY, U/C GRD. LOCK OVERRIDE, PULL UP. It remains up until returned manually to the down position.

(c) Post-Mod. 1344, a window is fitted in the port side of the fuselage, just below the rear end of the hood, enabling the pilot to check (on the ground) that the groundlock hook is engaged with a pin, thus locking the selector lever.

> NOTE.—It is possible for the undercarriage to be locked down but the groundlock not engaged, if the selector lever is not fully down on the stop.

(d) *Emergency operation*

If hydraulic pressure fails, the undercarriage may be raised or lowered by operating the handpump after making the normal selection.

21. Undercarriage position indicators

(a) There is an undercarriage position indicator (58) (113) in each cockpit. When electrical power is available, the indicators operate as follows:—

Undercarriage locked down	.. 3 green lights
Undercarriage unlocked	.. 3 red lights
Undercarriage locked up	.. All lights out

(b) If any wheel is not locked down and either throttle is less than one-third open, a micro-switch (6) operates and warning lights (56) (112) to the right of the indicators come on. At throttle settings giving 12,500 r.p.m. (approx.) these lights will remain on continuously.

(c) When the nosewheel is down and locked, a small rod protrudes about one inch above the nose of the aircraft.

22. Flap controls and indicators

(a) *Normal operation*

The hydraulically-operated flaps are controlled by two interconnected levers (32) (103), one on the left of each instrument panel, marked UP–NEUTRAL–DOWN. Any intermediate flap angle may be selected by moving either lever to UP or DOWN as appropriate and, when the flaps have reached the desired position, returning the lever to NEUTRAL. The flap lever should be returned to NEUTRAL after each operation of the flaps.

(b) *Emergency operation*

If hydraulic pressure fails, the flaps may be raised or lowered by operating the handpump after making the normal selection.

(c) *Position indicator*

There is an electrically-operated position indicator (34) (105) in each cockpit.

23. Airbrake controls

(a) *Normal operation*

The hydraulically-operated airbrakes are controlled by two interconnected levers (17) (95), one on the port wall of each cockpit, below the throttle levers. The levers are marked OUT (aft)–AIRBRAKE–IN (forward). There is no position indicator, the brakes being visible from both cockpits.

(b) *Emergency operation*

If hydraulic pressure fails, the airbrakes may be operated by the handpump, after making the normal selection.

24. **Wheelbrakes**

The pneumatically-operated wheelbrakes are controlled by a lever on each control column. Differential braking is obtained by operating the rudder pedals with either of the levers pressed. The pressure at each wheel (120 lb./sq. in. max.) may be read on the triple-pressure gauge (8) on the port wall of the front cockpit. A parking catch is incorporated on the front control column and, pre-Mod. 813, on the control column in the rear cockpit.

Page 21
Para.
5(a)
A.L.2.

25. **Flight instruments**

(a) *A.S.I. and associated instruments*

(i) A combined pitot-static pressure head on the port wing supplies pressure for the operation of the A.S.I.s, V.S.I.s and altimeters.

(ii) Mod. N.1878 introduces an altimeter Mk. 19B. This instrument has a striped sector which shows when the aircraft altitude is below 16,000 feet.

(iii) The pressure head can be electrically heated, the supply being controlled by a switch (61) on the starboard switch panel in the front cockpit.

(b) *Compasses and direction indicators*

(i) *Remote-indicating compasses*

Pre-Mod. 1177, there is a remote-indicating compass in the front cockpit, the electrical supply being controlled by a switch on the starboard switch panel in the front cockpit.

(ii) *Mk. 4F compass*

Mod. 1177 introduces a Mk. 4F compass (39) in the front cockpit, together with an A.C. voltmeter (53) below the main instrument panel and a corrector control box (126) in the rear cockpit. The electrical supply is controlled by a master switch (59) and D.C. (65) and A.C. (64) NORMAL-EMERGENCY switches (see para. 16(c)). When I.L.S. is installed, the Mk. 4F compass is replaced by a Mk. 4FT.

(iii) *Direction indicators*

Pre-Mod. 1177 there is a direction indicator in each cockpit. Post-Mod. 1177 there is one in the rear cockpit only. The gyros for these indicators are vacuum-operated (see para. 15).

(iv) *E.2.A compasses*

There is an E.2.A compass in each cockpit, to the right of the instrument panel. The one (42) in the front cockpit is below the panel coaming, while that (110) in the rear is at the bottom of the panel.

(c) *Artificial horizons*

The gyros for the artificial horizons are vacuum-operated (see para. 15).

(d) *Turn and slip indicators*

Pre-Mod. 1228 the turn and slip indicators are operated by the vacuum system. Post-Mod. 1228 the instruments are electrically operated by D.C. and have OFF flags to indicate power failure.

(e) There are clocks (46) (114) in both the front and rear cockpits (post-Mod. 1844 rear cockpit) to the right of the instrument panels.

(f) *I.L.S. and Zero reader*

Mod. 1636 introduces I.L.S., Zero reader and a Mk. 4FT compass. The gunsight is removed. In the front cockpit are the gyro unit, voltmeter, control panel, course selector, horizon gyro unit, Zero reader indicator, I.L.S. indicator, control unit, pushbutton and switch, and an amber lamp. In the rear cockpit are a Zero reader indicator, I.L.S. indicator, flight computer and repeater.

(g) An accelerometer may be fitted in place of the gunsight.

(h) Mod. 1891 introduces an altitude warning indicator. An amber flashing warning light in the centre of each instrument panel is barometrically controlled to come on when the aircraft descends below 11,061 ± 400 feet. Dimming facilities are available. Press-to-test and cancellation buttons are provided and are only operative below the set altitude.

COCKPIT EQUIPMENT

26. Cockpit entry

Footsteps and handholds on the port side of the front fuselage give access to both cockpits. Retractable foot-steps are lowered by pulling their external handles and are retracted by a control just forward of the front footstep. They can only be operated from outside the aircraft.

27. Hood operation

(a) When the hood is closed, it is supported and locked by catches on the port and starboard sides and by two hooks at the rear. When the hood is opened, the rear hooks and port catches are released, allowing the hood to swing upwards about the starboard catches until a folding strut (midway between the two cockpits) straightens and locks. A tie-rod then locks the hood-opening handles and port catches in the open position. When the hood is closed, the sleeve on the folding strut below the joint is pushed down and the strut folded downwards. The hood is then pulled down until it seats on the decking and is locked by means of the internal or external handle. *The external handle is for servicing only, as its action is not sufficiently positive to ensure absolute locking of the hood.*

(b) (i) There is a hood opening handle (11) (93) on the port wall of each cockpit. When either handle is pushed forward, it releases the port catches and rear hooks. While the hood is being opened, the handles are free to move. They should never be pulled back into the closed position before the hood is closed again, as the weight of the hood coming down might damage the rear hooks, which would be in the locked position. Post-Mod. 815, the hood cannot be closed with the handles in the locked position.

(ii) An external opening handle is provided. To open the hood, the catch at the rear of the handle is pushed forward and the rear end of the lever is pulled out and pushed upwards.

(iii) When the hood is locked, both internal handles should be one-quarter inch from the fully-down position on the guard (92). (This space is to ensure that the mechanism has been moved over the dead-centre position before reaching the stops; if there is no space, the locking mechanism should be checked.)

(iv) The toggle handles (77) (120), on the starboard wall of each cockpit. release the rear hooks when pulled forward. The handles should only be used when the hood-opening handle fails to release the rear hooks An external release, similar in action, is on the port side of the fuselage, under a fabric patch.

(c)
   (i) A hood locking indicator (73) is on the starboard wall of the front cockpit. When the starboard catches are correctly engaged, the two white pointers on the indicator are in line. Unless the pointers are exactly in line, there is a chance of the starboard catches releasing in flight.

   (ii) Mod 40/FC introduces a spring-loaded pushbutton on the port wall of the front cockpit; it can only be depressed when the operating handle is in the correct locked position.

   (iii) Mod 1787 introduces a hood warning light in each cockpit: in the front cockpit it is above the windscreen to port of centre and in the rear cockpit (106) it is at the forward end of the port wall. The lights come on when the operating handle is not in the fully closed position.

(d) *Hood jettison*

   (i) A hood jettison handle (44), (108) is on the starboard side of each instrument panel. When either handle is pulled out to its full extent (10-12 ins) both port and starboard catches are released, allowing the hood to be swung up and back about the rear hooks, from which it disengages at an angle of about 40°. The handle should be held fully out until the hood clears the aircraft, to prevent premature opening of the rear hooks.

   (ii) It is possible that, in a flat spin, there will not be enough aerodynamic lift to release the hood from the rear hooks. In this case, it is recommended that the ground jettison procedure is used (see Part IV Para 86(b)).

28. **Seats and harness**

Both seats can be adjusted for height by levers on the starboard side of the seats. The lean-forward releases (74) (121) for the Z-type harness are on the starboard wall of each cockpit. A dinghy guard is fitted to the front of each seat.

29. **Cockpit heating**

(a) The cockpits are supplied with warm air from the engines and the heating in both cockpits is controlled by a lever (71) on the starboard wall of the front cockpit. The forward position of the lever is the OFF position and the lever is moved progressively backwards to ON to increase the heat in both cockpits. This control should be OFF during take-off, to prevent fumes entering the cockpit.

(b) A ventilation control (14) on the port wall of the front cockpit operates a cold-air intake valve and a stale-air outlet valve. The control is pulled back progressively to open both valves.

30. Windscreen de-icing and de-misting

(a) A de-icing spray on to the windscreen centre panel is supplied from a four-pint tank in the nose and controlled by a handpump (82) on the front cockpit floor, to the right of the seat. The pump incorporates a flow control.

(b) Post-Mod. 1413 the centre and starboard windscreen panels are electrically heated to prevent misting up. The controlling switch (75) is on the starboard wall of the front cockpit.

31. Windscreen wiper

Mod. 1862 introduces a hydraulically-operated windscreen wiper. The control is on the port wall of the front cockpit, in the place normally occupied by the camera footage indicator (5); it is marked RUN–OFF–PARK. When not in use in the air the control should be left at PARK.

32. Direct-vision panels

There are two direct-vision panels, one in the windscreen port side panel and the other in the port side of the hood in the rear cockpit. Each panel is opened by unscrewing the appropriate handwheel ((13) front cockpit); the panel is then swung inwards until the handwheel is engaged by the catch.

33. Oxygen system

(a) Oxygen is carried in four bottles on the port side of the rear cockpit. In each cockpit the high-pressure supply is taken to a regulator (47) (109), on the starboard side of the instrument panel, which allows selection of NORMAL, HIGH or EMERGENCY flow. The oxygen tube (84) (127) and socket are to starboard of each seat. There is a line valve on the inter-cockpit decking which should be wire-locked in the ON position.

(b) Pre-Mod. 1239, the regulators are Mk. 11C. Post-Mod. 1239, the regulator in the front cockpit is a Mk. 11D and that in the rear a Mk. 11E. In this case, the supply to the rear regulator is taken from the front regulator.

34. **Internal lighting**

(a) *Front cockpit*

The instrument panel and starboard switch panel are illuminated by red and u/v lamps, controlled by dimmer switches (50) (57) at the base of the instrument panel. The auxiliary lights are controlled by a dimmer switch (16) under the port decking.

(b) *Rear cockpit*

The instrument panel is illuminated by red and u/v lamps, controlled by dimmers (116) (115) on the starboard wall. The auxiliary lamps are controlled by a dimmer switch (101) beside the trimmer controls, on the port side. When a j.p.t. gauge is fitted, its lamp is controlled by a dimmer switch (107) on the left of the instrument panel.

(c) *Emergency lighting*

Each cockpit has a separate alkaline battery to supply the emergency lamps. The lamp in the front cockpit is controlled by a dimmer switch (76) on the starboard wall, while that for the rear cockpit is controlled by a dimmer switch (117) beside the red lamps dimmer switch.

35. **External lighting**

(a) All external lighting is controlled from the front cockpit only.

(b) The switches for the navigation lights (62), the resin lights (60), the downward identification colour selector (66) and the downward identification morsing pushbutton (67) are grouped together on the starboard switch panel.

(c) The landing lamp under the port wing is controlled by an OFF–LOW–HIGH switch (18) on the cockpit port shelf. When switched to LOW, the motor moves the lamp to the half-extended position and switches on the lamp. When switched to HIGH, the lamp extends fully. The lamp should not be extended at speeds above 175 knots.

OPERATIONAL CONTROLS

36. **Radio and radar equipment**

(a) There are twin ten-channel V.H.F. sets in the rear fuselage. The controllers (23) (2) and the change-over switch (10) are on the port wall of the front cockpit. The sets may be any of the following:—

> TR 1934/35, TR 1934/36, two TR 1936, TR 1985/86, TR 1985/87.

(b) Intercomm. is through the amplification stage of the VHF.

(c) Mod N.1870 introduces UHF and standby for Naval aircraft. The controller is in the position normally occupied by the gunsight and the aerial changeover switch, UHF/Standby switch, standby channel changeover switch and standby power changeover switch are grouped together on the starboard wall of the front cockpit. The emergency power for the standby is provided by a 7 amp. hr. battery in the rear fuselage. The lighting for the controller is operated by a dimmer switch on the starboard side of the cockpit.

(d) There is a mic/tel socket (81, 118) on the starboard wall of each cockpit. Quick-release sockets may be fitted to the sides of the seats. There is a press-to-transmit switch (48) on top of each control column, with a spring-loaded mute switch beside it; these may be moved to the inboard throttle lever in the front cockpit. Post-Mod. 812, the MUTE–NORMAL switch (119) in the rear cockpit is on the starboard wall and is not spring-loaded. On Naval aircraft, the VHF/UHF cannot be turned on until the oxygen is on. A switch marked MIC by the oxygen regulator, must be put on after turning on the oxygen.

(e) The IFF controller (72) is on the starboard wall of the front cockpit, together with the F (70) and D (69) switches. G–MANUAL, G–AUTO and G–D switches (9) are on the port wall of the front cockpit.

(f) SRIM 2471 introduces Rebecca Mk. 8. The indicator and controller are on the starboard side of the cockpit.

(g) SRIM 2679 introduces UHF in addition to VHF for R.A.F. aircraft. The controller takes the place of the gyro gunsight. The VHF/UHF change-over switch and the tone control are beside the existing VHF switches. Lighting of the controller is operated by a dimmer switch on the cockpit starboard wall. UHF standby is not provided.

37. Gyro gunsight

An electrically-operated gyro gunsight Mk. 4B may be above the instrument panel in the front cockpit. The gunsight selector-dimmer control (68) is on the starboard side of the cockpit, the master circuit breaker (41) is to the right of the gunsight and the ranging control is incorporated in the port engine throttle lever. Spare filaments (37) are stowed to the left of the sight. The retraction control switch (38) is to the left of the gunsight and, in the event of electrical failure, it is possible to lower the gunsight by pushing in a manual control (40) below the sight. This control should be used only in an emergency, since servicing will be necessary before the sight can be used again. The sight is automatically lowered when the hood is jettisoned.

38. Cameras

(a) *G.45B camera*

A G.45B camera in the fuselage nose is controlled by the master switch (63) on the starboard switch panel. The camera operates with the recorder camera and the footage indicator mounting (5) is on the port wall of the front cockpit.

(b) *Recorder camera*

The recorder camera above the gunsight is operated by pressing any part of the button (49) on the control column in the front cockpit. The gunsight master circuit breaker must also be on. The button on the control column in the rear cockpit is inoperative.

39. Target-towing gear

When target-towing gear is installed, the hook is incorporated in the ventral tank and the target is released by switching ON the camera master switch and pressing the camera pushbutton on the control column. A micro-switch in the hook mechanism automatically releases the target if it rides dangerously high, provided that the undercarriage is up. If the normal release mechanism fails to operate, the target may be released by jettisoning the ventral tank.

40. Miscellaneous equipment

A map case (22) (125) is provided in each cockpit. In the rear cockpit there is a stowage (96) for I.F.P.E. goggles (not normally used).

EMERGENCY EQUIPMENT

41. Hand fire extinguishers

Two hand fire extinguishers are on the decking between the two cockpits.

42. **First aid**

Two first-aid packs are on the decking between the two cockpits.

43. **Dinghies**

Each pilot carries his own K-type dinghy.

44. **Crowbars**

A light crowbar is clipped to the port side of each pilot's seat.

PART II
LIMITATIONS

NOTE.—The limitations quoted in Part II are mandatory and are not to be exceeded. The contents of Parts III and IV are mainly advisory but instructions containing the word " must " are to be regarded as mandatory.

45. Engine data—Derwent 8 and 9

The principal engine limitations are as follows:—

Power rating	Time limit	Maximum permissible r.p.m.	Maximum permissible j.p.t. ($^{\circ}$C.)
Take-off, climbing and operational necessity	15 mins.	14,700 ± 100*	680 (700 above 20,000 ft. for ten minutes)
Maximum continuous	No limit	14,100	630
Idling on ground (throttle closed)	No limit	3,300–3,700	500

* A governor on the engine-driven fuel pump restricts the r.p.m. to 14,550 at sea level but with the throttle fully open r.p.m. may increase progressively to 14,700 with height.

Oil pressures

Minimum for idling	5 lb./sq. in.
Minimum at 14,100 r.p.m. and above ..	30 lb./sq. in.

46. Flying limitations

(a) The aircraft cockpits are not pressurised and the maximum permissible altitude is 30,000 feet for any period exceeding 10 minutes and the absolute maximum is 35,000 feet for not more than 10 minutes. Spinning and stall turns are prohibited. Maximum rate turns and other manoeuvres involving high G are prohibited when there is fuel in the ventral tank.

WARNING.—It is possible to cause excessive G loading by flying the aircraft out of trim. See also Para 64(d).

PART II—LIMITATIONS

(b) *Maximum speeds and mach number limitations* (see also para. 64)

    (i) Clean aircraft

Sea level to 1,000 ft.	510 knots
1,000 ft. to 20,000 ft.	0.76M
Above 20,000 ft.	0.78M

    (ii) For raising and lowering undercarriage — 175 knots

    With undercarriage locked down — 195 knots

    (iii) For lowering flaps — No limit (relief valve in system prevents excessive strain)

    For flaps fully down — 150 knots

    (iv) For opening airbrakes — No limit

    (v) For opening direct-vision panels — 200 knots

    (vi) With ventral drop tank:—

    At all heights — As for clean aircraft

    For jettisoning the ventral drop tank — 345 knots

    NOTE.—Aerobatics are prohibited when carrying the ventral drop tank unless the tank is empty.

    (vii) With 2 × 100-gallon wing drop tanks:—

    When wing drop tanks are fitted, aerobatics are prohibited and speed is restricted as follows:—

    From strength considerations — 435 knots

    From control considerations — 0.70M up to 20,000 ft. 0.72M from 20,000 ft. to 30,000 ft. 0.74M above 30,000 ft

Wing drop tanks may be jettisoned at any speed up to 260 knots.

Unless Mod. 1610 has been incorporated, the rate of descent when carrying wing drop tanks should be restricted to 5,000 ft./min., otherwise, due to the insufficient equalising of pressures, there is a risk of tank collapse. Slight damage to the tank nose increases the risk of tank collapse; however, if collapse should occur, there is no danger or change of trim.

(c) *Ground attack*

If the angle of dive near the ground exceeds 10°, the maximum permissible speed is 400 knots. The maximum permitted angle of dive in ground attack is 45°.

(d) *Maximum weights*

For take-off and gentle manoeuvres .. .. 18,800 lb.
For all permitted forms of flying and landing .. 14,700 lb.

At heavy loads the aircraft should be operated only from runways and rapid turns on the ground must be avoided.

(e) *C.G. limitations*

Forward limit .. .. .. 2.6 ins. forward of datum
Aft limit .. .. 2.3 ins. aft of datum

(f) *G limitations*

Any configuration .. .. — 3G
Clean or with empty ventral tank .. .. .. + 6G
Wing drop tanks, full or empty, ventral tank unless
   empty .. .. + 4G

(g) *Aircraft approach limitations (A.A.L.s)*

(i) The aircraft approach limitations are as follows:—
   Precision radar (G.C.A.) .. 200 ft. above airfield level
   Manual I.L.S. .. .. 300 ft. above airfield level
   I.L.S./Zero Reader ·· ·· 200 ft. above airfield level

(ii) The above A.A.L.s are subject to the following conditions:—
   1. The pilot must be qualified on the aircraft and be in current practice on the aircraft and the aid used.
   2. Normally, a standard cross-bar airfield approach lighting system should be in operation and at least two crossbars should be visible from break-off altitude. Where a standard system is not available, the approach must not be continued unless the runway threshold is visible from break-off altitude.
   3. The controller must announce to the pilot arrival at break-off altitude.

PART III

HANDLING

MANAGEMENT OF SYSTEMS

47. Management of the fuel system

(a) The L.P. pumps and the L.P. and H.P. cocks must be on for starting and for all normal forms of flight.

(b) Use fuel from the wing drop tanks first, if these are fitted. When WING ON is selected, the fuel transfer warning light comes on momentarily and then goes out, indicating that fuel is being transferred; the fuel gauge readings should remain constant. As soon as the light comes on again (or when the gauge readings start decreasing) select BELLY ON; the warning light will go out. The gauge readings will drop to approximately 125/125 gallons and then remain constant. When the ventral tank stops feeding, the total remaining fuel is indicated on the contents gauges; the transfer warning light will come on and remain on. On pre-Mod. 1841 aircraft, return the fuel transfer handle to OFF; the light will then go out.

(c) Freezing of the ventral tank inward vent valve may occur whilst flying at altitude with wing tanks selected. This will prevent the ventral tank from feeding when it is ultimately selected. Freezing of this valve can be avoided by carrying out the following drill:—

1. After starting, check the ventral tank flow.

2. Take-off with wing tanks selected.

3. Until the wing tanks are empty, make frequent selections of the ventral tank.

(d) *Use of the balance cock*

   (i) *When flying on two engines.* Any out-of-balance state can be quickly corrected by use of the balance cock and, if necessary, altering the attitude of the aircraft. Unequal balancing in the main tanks will occur, particularly as a result of steep climbs or descents; this is aggravated if the balance cock is open. Under these conditions, if the fuel level is low, the fuel outlet of the uppermost tank may be uncovered, causing failure of the engine feeding from it. Normally, the balance cock should be closed.

(ii) *When flying on one engine.* If a choice is possible, the starboard engine should be used when flying on one engine to conserve fuel, for the following reasons.

    1. In level flight at range speed, the main fuel tank is in a slightly nose-up attitude; therefore, with the balance cock open, fuel will tend to flow into the rear compartment, from which the starboard engine is supplied.

    2. The hydraulic pump is driven by the starboard engine.

The following table gives a guide to the use of the balance cock during single-engined flying:—

Condition of flight	Engine in use	Position of balance cock
Climb	Port	Closed
	Starboard	Open
Descent	Port	Open
	Starboard	Closed
Level	Port	Closed only if I.A.S. below approx. 250 knots
	Starboard	Open

    NOTE.—With the live engine at or near full power, its fuel demand will normally be greater than the flow through the balance cock. Because of this, unless the fuel exceeds 35 gallons in the compartment not directly feeding the engine, it is possible to empty the other compartment.

(e) *Booster pump failure*

Should an L.P. pump fail during flight, it will be impossible to obtain maximum r.p.m. at high altitudes, or at high speed at low altitudes, on the engine concerned. In addition, the engines may flame out due to fuel aeration. The

L.P. booster pump circuit breakers may cancel in flight; should this happen, they should be pressed in again. Electrical failure should not be assumed unless the button will not stay in.

48. Engine handling

(a) *Starting*

Either engine may be started first but the order in which the engines are started should be alternated on each sortie in order to check the functioning of the vacuum pumps.

(b) *Take-off and climb*

With the throttles fully open, r.p.m. at take-off are governed at 14,550. During the climb, r.p.m. will gradually increase to 14,700. A close watch should be kept on the j.p.t., which must not be allowed to exceed 680 C. (700 C. above 20,000 feet for 10 minutes). If climbing at less than full power, it will be necessary to reduce the throttle setting to maintain the selected r.p.m. If surging occurs—recognised by a muffled detonation in the engine—the r.p.m. should be reduced until it ceases.

(c) *General*

(i) There is little or no risk of flame extinction at altitude, providing throttle movements are made smoothly and not too rapidly. Coarse opening of the throttles will, however, cause surging and excessive j.p.t.

(ii) The throttles may be closed fully at any altitude. During descent, the throttles will have to be opened progressively to maintain a selected r.p.m.

(iii) False indication of oil pressure failure may be given whilst at height. A descent to a lower altitude should return the oil pressure indication to normal.

(iv) After periods at height, the H.P. cock may freeze in the on position. It will free during the descent.

STARTING, TAXYING AND TAKE-OFF

Page 35
Para.
49 (a)
A.L.4.

STARTING, TAXYING AND TAKE-OFF

49. External checks

(a) Before starting the external checks, look inside the cockpit and ensure that the undercarriage selector lever is fully DOWN and that there is sufficient brake pressure and oxygen. Check the hydraulic pressure at 800 lb./sq. in. minimum on the gauge on the starboard side of the rear cockpit floor. Ensure that the control locks are removed.

(b) The outside of the aircraft should be checked systematically for obvious signs of damage and security of panels, doors and mudguards. The main wheel oleos should be checked for equal extension, the tyres for cuts and creep and the brake leads for damage. Check that, with the elevator in the neutral position, the horn balances are in line.

(c) The following specific checks should also be made:—

Starboard undercarriage bay	Isolating switches on
Starboard wing	Engine intake free from obstruction
	Jet pipe for wrinkling and turbine for damage
Port wing	As for starboard wing
	Pressure-head cover off
Port undercarriage bay	As for starboard undercarriage bay
Fuselage tank panels	Secure
Undercarriage ground-lock hook behind window	Engaged with pin

50. Internal checks

(a) Before entering the cockpit, check that an external battery is plugged in and switched on and the hood external handle is in its housing.

(b) If the aircraft is being flown solo, make the following checks in the rear cockpit:—

No loose equipment	
Straps done up	
Mute switch (119)	NORMAL
Crowbar	Stowed
Oxygen tube (127)	Stowed
Undercarriage ground lock override (91)	Down
Oxygen (109)	OFF
Hood jettison control (108)	In
Hood rear-hooks toggle (120)	Back
Hood D.V. panel	Closed

| Direction indicator | Caged |
| Cockpit lighting | Off |

(c) *Inter-cockpit decking checks*

Oxygen cock	On
First-aid kits	Stowed
Fire extinguishers	In position and secure
Internal control locks	Correctly stowed
Picketing rings	Stowed and wired in
Asbestos gloves	Secure
Silica-gel crystals	Condition

(d) *Front cockpit checks*

Strap in and ensure R/T lead is plugged in. Before closing hood, check the hood warning light on (post-Mod. 1787). Instruct the groundcrew to close the hood and retract the footsteps. Adjust the seat and rudder pedals and check the flying controls for full, free and correct movement. Apply the parking brake and check total brake pressure (200 lb./sq. in. minimum). Test the brakes for maximum delivery to each wheel (120 lb./sq. in.) and equal pressure with the rudder bar central. Then check:—

Hood handle (11)	$\frac{1}{4}$ inch from fully down position
Hood plunger (post-F.C. Mod. 40)	Moves freely
Direct vision panel (13)	Closed
Hood warning light	Out
Hood jettison pointers (73)	Exactly in line
Hood rear hook toggle (77)	Fully back
Harness	Adjusted
Crowbar	Stowed
H.P. cock (3)	OFF
L.P. cock (1)	ON
Balance cock (29)	Free movement. Leave down
Trimmers (25, 28)	Full, correct and free movement. Set rudder neutral, elevator one div. NOSE-UP.

Windscreen wiper (post-Mod. 1862)	PARK
VHF/UHF	All off, select as required
Airbrakes (17)	Check wings clear and select OUT and IN
Throttles (12)	Full and free movement. SHUT
Ventilation lever (14)	As required
Undercarriage ground-lock override (15)	Down
Fuel transfer lever (35)	Fully in. Selection as required. Warning light on if selected to WING ON or BELLY ON
Flaps (32)	Check operation. Exhaust the accumulator and use the handpump. Select UP
Undercarriage selector lever (31)	DOWN
Undercarriage position indicator (58)	Three green lights. Check bulb changeover. Warning light (56) out.
Landing lamp selector switch (18)	OFF
L.P. pumps (19)	On individually. Check aurally
Fire warning lights (36, 43)	Out
Accelerometer (if fitted)	Set to +1G.
Gyro gunsight	Retracted
Instruments	Check serviceability. Set altimeter
Oil pressure gauges (51, 55)	Serviceability
Fuel gauges (52, 54)	Contents
Generator warning lights (post-Mod. 1225)	On
E.2A standby compass (42)	Heading
Hood jettison handle (44)	Fully in
Fire extinguisher buttons (45)	Covered
R.p.m. and j.p.t. gauges	Serviceability

Oxygen (47)	Fully ON. HIGH flow selected. Emergency OFF. Contents. Connections and flow check. On Naval aircraft MIC switch on
AC/DC switches (post-Mod. 1177) (64) (65)	NORMAL
Pressure-head heater (61)	ON
Compass switch (59)	ON. Check Mk. 4F compass starts and voltmeter indicates in white sector Pre-mod. 1177, align D.I. with compass
Other switches	As required
Clock	Correct and working
Cockpit heater (71)	OFF
Harness release (74)	Test
Windscreen heater switch	As required
Emergency lamp dimmer switch (76)	OFF
R/T lead	Plugged in
H.P. cock (79)	OFF
L.P. cock (80)	ON
Windscreen de-icer pump plunger (82)	Down and locked
Wing drop tanks jettison lever (83)	Fully forward
Hydraulic emergency hand-pump (85)	Retracted
Internal and external lights	As required

51. Starting the engines

(a) *Pre-starting checks*

Confirm the following:—

L.P. cocks	On
H.P. cocks	OFF
Throttles	Closed
L.P. pumps	On (circuit breakers in)

(b) *Starting*

> NOTE.—1. The engines may be started on the aircraft batteries. This, however, imposes an undesirable load on them and is not recommended. An external battery should normally be used for starting.
>
> 2. If an engine fails to start after two attempts, the cause should be investigated before making further attempts.

(c) Press the starter pushbutton of the selected engine and release it after two seconds.

(d) When the undercarriage lights dim, move the H.P. cock to the half-open position. When r.p.m. increase, the cock should be moved slowly to the fully-open position. The engine should accelerate to idling speed with the throttle closed. The exhaust temperature may momentarily exceed the idling limit but it should settle down to not more than 500°C. The throttle must not be opened before idling r.p.m. are attained.

(e) If the H.P. cock is moved too quickly from the half to the fully-open position, resonance and overheating may occur. If excessive exhaust temperatures and resonance persist, close the H.P. cock to stop the engine. Excess fuel must have drained off before another start is attempted.

(f) If an engine fails to light-up, proceed as follows:—

  (i) Turn off the H.P. cock.

  (ii) Have the appropriate isolating switches set off.

  (iii) Ensure that the impeller has stopped turning. Wait until the fuel has stopped draining from the nacelle and then dry out the engine by carrying out the starting cycle with the H.P. cock in the off position.

  (iv) When the impeller has again stopped turning, have the ground crew remove any surplus fuel from the jet pipe.

  (v) Have the isolating switches set on.

  (vi) Start the engine as in sub-paras (b) to (d) above.

WARNING.—*Derwent 9 engines.* When charged, the capacitor in the high energy ignition unit possesses a lethal voltage. The unit must be isolated and at least one minute allowed to elapse, before the ground crew make adjustments in the vicinity of the unit.

(g) Simultaneous starting of both engines is not permissible but, when for operational purposes it is necessary to start the engines with minimum delay, the starter pushbutton of the second engine may be pressed not less than 5 seconds after the starter pushbutton for the first engine has been released. The starting sequence then becomes:—

   (i) Starboard engine starter button—press and release.

   (ii) When the starboard engine r.p.m. reach 800–1,000, turn the H.P. cock half on.

   (iii) Port engine starter button—press and release.

   (iv) When the port engine r.p.m. reach 800–1,000, turn the H.P. cock half on.

   (v) Ease both H.P. cocks to the fully-on position.

(h) If, while starting the starboard engine, a severe hammering is heard, it is due to air in the hydraulic system, and should disappear when one of the hydraulic services is operated.

(j) When both engines are running at idling r.p.m. and not before, have the ground/flight switch placed to FLIGHT and the ground starter battery disconnected.

52. Checks after starting

Engine fire-warning lights	Out
Idling r.p.m.	3,300 to 3,700 r.p.m.
J.p.t.	Max. 500° C.
Oil pressure	5 lb./sq. in. (min.)
Pneumatic pressure	200 lb./sq. in. (minimum for taxying) and increasing (450 lb./sq. in. max.) Pressure at each wheel 120 lb./sq. in.
V.H.F.	Channel and set selected

Flight instruments	Check and set
Generator warning lights (if fitted)	Out at approx. 5,000 r.p.m.
Flaps	Up. Return lever to NEUTRAL
Airbrakes	Test—IN

53. Taxying

(a) Considerable power is necessary to move forward initially. At normal taxying speeds, gentle turns can be made by using the engines but a combination of engine and brake is required to make small-radius turns.

(b) The throttles should not be opened rapidly, as this may cause over-fuelling and excessive j.p.t.

(c) When taxying, fuel consumption is about 1 gallon per minute, for each engine, at idling r.p.m.

54. Checks before take-off

Page 42 Para. 54 A.L.2.

Trim	Elevator: 1 div. nose-up Rudder: neutral
Airbrakes	IN
Fuel	Contents L.P. and H.P. cocks fully on. L.P. pumps on. Ventral or wing drop tanks selected. Balance cock closed
Flaps	Up or one-third down
Instruments	Artificial horizon erected. Mk. 4F compass on and synchronised. Turn and slip OFF flag not visible, correct functioning. Pressure-head heater ON
Oxygen	Fully ON. Contents. Emergency OFF. HIGH flow. Connections checked
Hood	Closed. Pointers exactly in line. Handles ¼ in. from horizontal position. Warning lights out (post-Mod. 1787)

Heating	OFF
Harness	Tight and locked
Controls	Full and correct movement

55 Take-off

Page 43
Para.
55 (a) (b)
A.L.4.
(a) Align the aircraft for take-off, with the nosewheel straight, and apply the brakes. Open the throttles smoothly, synchronising r.p.m. At 11,000 r.p.m. check that the brakes are holding and oil pressures. When the aircraft begins to creep forward, release the brakes and open the throttles fully. At full power, check the fire warning lights out, r.p.m. 14,550 and j.p.t. below 680°C.

(b) The rudder becomes effective at about 70 knots. There is no tendency for the aircraft to swing.

(c) Ease the nosewheel off the ground at 80–90 knots. Care should be taken not to get the nosewheel too high; coarse backward movement of the control column may cause the tail to touch the ground. The aircraft does not unstick cleanly and should be flown off at 120 to 130 knots, depending on load.

(d) Safety speed is 160 knots (165 knots on aircraft with large air intakes).

56. Checks after take-off

(a) When comfortably airborne, brake the wheels and then retract the undercarriage. To avoid risk of damaging the undercarriage, ensure that retraction is completed before reaching 175 knots.

(b) When safety speed has been attained, raise the flaps (if used), returning the selector to NEUTRAL. If drop tanks are fitted, check that the fuel transfer warning light is out, with either WING ON or BELLY ON selected. Select heater on if required.

Page 43
Para.
56(c)
(c) Unless it is necessary to clear obstacles, allow the speed to reach that recommended for climbing, before starting the climb. As the aircraft accelerates to climbing speed, there is a progressive nose-up change of trim.

HANDLING IN FLIGHT

57. **Climbing**

(a) Climb at full power within the j.p.t. limitations. The following climbing speeds are recommended with or without drop tanks fitted:—

> 300 knots until 0.63M is obtained, then 0.63M for the remainder of the climb.

(b) If maximum rate of climb is not essential, the climb may be made at 14,100 r.p.m., climbing at 280 knots until 0.63M is obtained and then using 0.63M for the remainder of the climb. Throttle adjustment will be necessary to maintain the power setting.

(c) Particular care must be taken to maintain the correct climbing speed at altitude, or a considerable increase in time to height will result.

57A. **Compressor surge**

Compressor surge can occur at high altitude and low I.A.S. The indications of surge are a muffled thumping, fluctuating r.p.m. and a rise in j.p.t. If surging occurs, throttle back the affected engine and increase speed.

58. **General flying**

(a) *Flying controls*

(i) *Ailerons.* On aircraft fitted with geared tabs only, the aileron feel becomes progressively heavier with increasing airspeeds and mach numbers. When spring tabs are fitted, the ailerons are light and pleasant but become heavier with increase in mach number. Large control column movements are required for effective control at low I.A.S.

(ii) *Elevator.* The elevators are effective and sensitive throughout the speed range. At low airspeeds they become less sensitive but are effective down to the stall.

(iii) *Rudder.* Control forces on the rudder become very heavy at high speeds and with large deflections at low speeds. Sustained asymmetric flight at low I.A.S. is very tiring.

(b) *Trimming controls*

> NOTE.—The elevator and rudder trimmers may ice up at height. Exercising the trimmers at regular intervals will help to prevent this happening. If necessary, reduce height until they can be freed.

(i) *Elevator trimmer.* The elevator trimmer is spongy in operation but very effective and the aircraft is easy to trim under most conditions of flight.

(ii) *Rudder trimmer.* The rudder trimmer is awkward to operate but is effective at high airspeeds.

(c) *Airbrakes*

The airbrakes are effective and will open fully at any speed. They make the aircraft less pleasant to control laterally. At speeds above 460 knots buffeting may be severe. When descending with airbrakes out, throttles closed, the rate of descent is high and altimeter errors will be considerable. Before closing the airbrakes, the angle of dive should be reduced.

(d) *Changes of trim*

Increase in power	Strong nose-up
Increase in speed	Strong nose-up
Operating flaps, under-carriage or airbrakes	Negligible

WARNING.—Because of the strong nose-up change of trim with increase in speed, elevator forces must be trimmed out up to the maximum speed attained and pull forces must be applied with care. It is possible to cause excessive G loading by flying the aircraft out of trim.

59. Endurance

If it becomes necessary to fly for best endurance, the following broad rules should be applied:—

(a) If at 25,000 feet or above, maintain height.

(b) Below 25,000 feet, with 300 gallons or more remaining, climb to 25,000 feet.

(c) Below 25,000 feet, with less than 300 gallons, use only one engine and do not climb.

60. Flying at reduced speed

Reduce speed to 170 knots. At this speed handling is easier if $\frac{1}{3}$ flap is used.

61. Flying in severe turbulence

The recommended speed for flying in conditions of severe turbulence is 215–225 knots up to the height at which this corresponds to 0.5M, i.e. up to about 20,000 ft. Above this height, 0.5M, should be maintained, providing this does not require an airspeed of less than 195 knots.

62. Stalling

(a) The appropriate stalling speeds (in knots), power off, are as follows:—

Aircraft configuration	With full internal fuel and two pilots	With full internal fuel, full ventral drop tank and two pilots	As in previous column, with full wing tanks
Undercarriage and flaps up	100	105	120
Undercarriage and flaps down	90	95	110

(b) Warning of the approach of the stall is given by:—

(i) Slight elevator buffeting some 10 knots before the stall, becoming more pronounced as the stall approaches.

(ii) Slight fore and aft pitching accompanied by vibration before the aircraft stalls.

At the stall, the nose drops gently and there is a slight tendency for either wing to drop, if the control column is held back. Recovery in all cases is straightforward.

(c) With the airbrakes out, the stalling speeds quoted above are increased by 2–3 knots and the buffeting before the stall is more pronounced.

(d) *G-stalling*

Onset of G-stalling is indicated by buffeting; continued rearward movement of the control column will cause more severe buffeting accompanied by a large loss of speed. Recovery is immediate when the back-pressure is released.

63. Spinning

(a)   Intentional spinning is prohibited.

(b)   The aircraft normally enters the spin by first rolling onto its back and then dropping its nose. Behaviour in the spin varies widely between aircraft and between spins on the same aircraft. Pitching, changes in rate of rotation and snatching of the control column from side to side are characteristics which can occur in varying degrees in various spins, although the control column normally snatches over to the inside of the spin.

(c)   Standard recovery action is effective but should not be taken while the aircraft is inverted or when the nose of the aircraft is above the horizon. Standard recovery action is:—

    (i)   Check throttles closed, then full opposite rodder followed by a short pause.

    (ii)   Ailerons neutral.

    (iii)   Stick steadily forward, to the fullest extent if necessary, until the spin stops.

    (iv)   Centralise all controls immediately.

    (v)   Recover to normal flight.

Prior to recovery, speeding up of rotation and steepening of attitude may be experienced. Heavy control forces must be expected; a two-handed force may be required to centralise the ailerons.

(d)   If standard recovery is not effective, the following supplementary actions are recommended:—

    (i)   Check that *full* opposite rudder is applied, that the stick has been moved *fully* forward and that the ailerons are neutral.

    (ii)   Check that the undercarriage and flaps are up.

    (iii)   Jettison any external stores.

    (iv)   Apply full in-spin aileron (in the direction of the spin). The recovery with in-spin aileron applied is likely to develop into a steep spiral with rapidly increasing speed unless the ailerons are centralised immediately the spin stops.

    (v)   Open up the inboard engine. Close the throttle immediately the spin stops, to avoid entering a spin in the opposite direction.

(e)   To recover from an inverted spin:

    (i)   Throttles closed, then full rudder (in the direction of roll) to oppose yaw.

    (ii)   Ailerons neutral.

    (iii)   Stick steadily back until the spin stops, then

    (iv)   Centralise all controls immediately.

    (v)   Stop any roll and recover from the dive.

(f)   If in doubt as to whether a spin is inverted or normal:

    (i)   Centralise the controls in an attempt to recover in the incipient stage.

(ii) Should the spin continue, apply full pro-(normal) spin control and recover from the ensuing normal spin in the manner detailed in (c) above.

64. High-speed/flying

(a) The high mach number characteristics vary between individual aircraft and, generally speaking, become worse with age and deterioration of the external finish. The investigation of an aircraft's performance should be carried out progressively until its high-speed characteristics are established and the limitations quoted in para. 46 must not be exceeded intentionally.

(b) *Clean or with belly tank*

(i) Above 20,000 feet, there is a progressively strong nose-up trim change as speed is increased; elevator trim may be used to counteract this tendency up to 0·76M, after which the trimmer must not be used. Slight buffeting may occur between 0·76M and the speed limitation of 0·78M and, if the latter speed is inadvertently exceeded in a shallow dive, a heavy push-force is required; pitching, sometimes accompanied by slight snaking, may occur as the speed approaches 0·82M. On some aircraft, a nose-down change of trim may be felt at this speed but, in any case, all control forces are very high and wing dropping may occur. The application of positive G will produce these effects at lower mach numbers.

(ii) Severe vibration may be experienced on some aircraft below the mach number limitation. Should this happen, no attempt is to be made to fly at a higher mach number. If the mach number limitation of 0·78M is inadvertently exceeded in a diving turn, violent wing drop and complete loss of control may occur, particularly if high positive G is applied. The ensuing dive or spiral is likely to be very steep; buffeting and high stick loads will be experienced and a great loss of height is inevitable.

(iii) When manoeuvring at high mach numbers at or just above 20,000 feet particular care is needed.

(iv) Below 20,000 feet, the elevator trimmer may be used to counteract the strong nose-up trim change, up to the mach number limitation of 0·76M, after which it must not be used. At speeds in excess of 0·76M, the push force required makes it difficult to hold the aircraft in a dive. The aileron forces are very high on aircraft fitted with geared tabs but rather less when spring tabs are fitted.

(c) *Recovery from high-speed dives*

Select airbrakes out, close the throttles and gently ease the aircraft out of the dive; application of high positive G may delay the recovery. If the aircraft is out of control in a dive or spiral the controls will become effective as height is lost and mach number reduced; while in this out-of-control condition, it is particularly important to avoid high positive G until the mach number is reduced, because its application will nullify the beneficial effects of reducing mach number and prolong the uncontrollable dive. If the aircraft is inverted, recovery should be made by rolling rather than pulling through.

(d) *High-speed, low-level flying*

Page 49
Para
64(d)
A.L.1.

During high-speed runs near the ground, if the aircraft is out of trim (nose-up) and a rapid pull-out is made, elevator forces may suddenly lighten, inducing excessive G and resulting in structural damage. For this reason, stick forces *must* be trimmed out up to the maximum speed attained and great care must be exercised when applying pull forces. If the angle of dive near the ground is greater than 10°, the maximum permissible speed is 400 knots, at which speed more than 3 divisions of nose-down trim may be necessary. The angle of dive in ground attack must not exceed 45°.

65. Aerobatics

Until experience is gained, the following speeds in knots are recommended:—

Roll	..	..	..	..	.. 260–280
Loop	..	..	..	..	.. 370 380
Roll off the top	..	..	..	..	380 400
Climbing roll	..	..	..	..	390 plus
Stall turn	..	..	..	..	.. 280 300

In manœuvres in the looping plane, much height may be gained or lost and an ample margin must always be allowed for regaining normal flight. With experience, the above speeds can be reduced considerably.

NOTE.—The negative G traps in the main tanks ensure a supply of fuel for 15 seconds inverted flight; this must not be exceeded owing to the possibility of oil starvation. If a ventral drop tank is fitted, aerobatics must not be carried out unless it is empty. If wing drop tanks are fitted, aerobatics are prohibited.

66. Descent

(a) *Gliding*

Glide at 170 knots at all heights for maximum range. The distance covered in the glide will be approximately 2 miles per 1,000 feet of height lost.

(b) *Fast rate*

Page 49
Para.
66 (b)
A.L.4.

Set 10,000 r.p.m. (12,000 asymmetric), airbrakes out, and maintain 250 knots. The rate of descent will be 4,500 to 5,000 feet per minute.

(c) *Maximum rate*

Set throttles closed, airbrakes out, and maintain 0.7M until 350 knots is reached. The rate of descent will be approximately 15,000–20,000 ft. per minute. This descent can be used in cases of operational necessity; misting will be considerable at lower altitudes although the front windscreen will normally remain clear if a landing is made shortly after the descent. In any case, a reserve of fuel should be kept to give time to clear the misting. (See para. 58 (c).)

CIRCUIT PROCEDURE AND LANDING

67. **Circuit procedure**

NOTE.—At least 30 gallons of fuel per engine should be allowed for the circuit and landing.

(a) *Joining*

Reduce speed to 200 knots, select ¼ flap and set 10,500 r.p.m. to give a speed of 170 knots downwind.

(b) *Checks before landing*

Airbrakes	IN
Undercarriage	Down below 175 knots. Lever fully down
Fuel	Contents. Balance cock as required.
Flaps	As required
Harness	Tight and locked
Undercarriage indicator	Three green lights. Nose-wheel indicator showing
Brakes	Pressure. On, off, exhausted

NOTE.—If the aircraft is yawed at speeds below 170 knots with the airbrakes out, the nose may drop suddenly and the elevators become ineffective until the yaw is removed or the airbrakes retracted. The tendency is aggravated if a ventral tank is fitted. *Airbrakes should not be used at airspeeds below 170 knots at circuit height and should be in before the undercarriage is lowered.*

(c) *Approach and landing*

> NOTE.—The aircraft should not be landed with fuel in the ventral tank. The aircraft is at maximum landing weight with two pilots and full internal fuel and empty ventral and wing drop tanks.

(i) Turn across wind at 150 knots with approximately 8,500 r.p.m. selected. Lower full flap when required and start the final approach at 130 knots. To ensure immediate engine response, maintain at least 7,500 r.p.m. until certain of landing. Reduce speed progressively and cross the threshold at 110 knots. As the touch-down point is approached, the rate of descent should be checked and the aircraft flown gently onto the runway.

(ii) If the speed is allowed to fall off early in the approach, considerable power will be required to reduce the rate of sink.

(iii) If the windscreen wiper is being used, it may stop while full flap is being lowered.

(iv) The landing is straightforward and presents no difficulty.

(d) *Braking*

To achieve aerodynamic braking, hold the nosewheel off the runway but avoid an excessive nose-up attitude as the elevator remains effective down to about 80 knots and it is possible to strike the tail on the ground. When the nosewheel is on the gr und, apply brake progressively and continuously. If it is essential to obtain the shortest landing run, in cases of necessity, lower the nosewheel on to the runway shortly after touch-down and apply full brake, holding it on continuously. If landing on a wet runway, intermittent braking is more effective, as there is less likelihood of the wheels locking when using this method.

68. Going round again

(a) The power required and the fuel used depends on when the decision to go round again is taken. If the decision is made on the approach at approximately 300 feet and it is essential to conserve fuel, the use of 12,000 to 13,000 r.p.m. will give a satisfactory performance. Going round again under these conditions requires approximately 15 gallons of fuel. Going round again after touchdown is straightforward, using full power initially. The fuel consumption in this case is higher and approximately 30 gallons of fuel should be allowed for the complete circuit.

(b) In all cases:—

   (i) Open the throttles smoothly to give the required r.p.m.

   (ii) Raise the undercarriage and the flaps. There is no sink.

   (iii) After reaching 160 knots (165 knots on large air-intake aircraft) commence climbing.

69. Instrument approach

(a) The aircraft approach limitations (A.A.L.) are given in Part II.

Page 52 Para. 69 (b) (c) A.L.4. (b) The following speeds, flap settings and approximate power settings are recommended for use during instrument approaches with the undercarriage lowered:—

	R.p.m.	Flaps	Airspeed knots
Downwind	11,000	One quarter	150
Base leg	11,000	One quarter	145
Glide path	11,000	Full	130

(c) When making an asymmetric approach, the above r.p.m. settings should be increased by 2,000 r.p.m. (approx.) and ¼ flap should be used. It is recommended that the undercarriage is left up until approaching the glide path but it should be down and locked before starting the descent. Select full flap before reaching the runway threshold.

70. Flapless landing

Page 52 Para. 70 A.L.2. Maintain 150 knots crosswind, with throttles set to 7,500 r.p.m. On turning into wind, allow the speed to fall to 130 knots, maintaining this speed until near the runway threshold. Close the throttles and cross the threshold at 115–120 knots. The aircraft is in a nose-up attitude but vision from the front seat is hardly affected. Speed drops off slowly and the aircraft requires a long landing run. If necessary, the run may be reduced by turning off the H.P. cocks immediately after touchdown.

NOTE.—At speeds below 150 knots, the ailerons should be used with care as excessive yaw may develop, which could lead to a temporary loss of control and a large loss of height during the final turn.

PART III—HANDLING

71. Crosswind landing

The "crab" technique should be used for crosswind landings. Use the normal threshold speeds. Kick off the drift and place the aircraft firmly on the ground. Some differential braking may be required to keep straight.

72. Checks after landing

Brakes	Pressure sufficient for taxying (200lb./sq. in. min. in system)
Flaps	Up. Lever NEUTRAL
Pressure-head heater	Off

73. Shut-down procedure

(a) After allowing r.p.m. and j.p.t. to stabilise, stop the engines by turning off the H.P. cocks.

(b) After the engines have stopped, check:—

Heater	Off
Compass	Off
Windscreen heater	Off
Oxygen	Off
L.P. pumps	Off
Balance cock	Up
V.H.F. or U.H.F.	Off
Chocks	In position
Brakes	Off
Ground/flight switch	GROUND

NOTE.—It is important to ensure that the engines have stopped turning before putting the ground/flight switch to GROUND; if this is not done, the j.p.t. gauges may be damaged.

ASYMMETRIC FLYING

73A. Asymmetric flying, general

(a) The hydraulic pump is on the starboard engine and, with this engine windmilling, there should be enough pressure to operate each service once, one way only. If the engine has seized, however, or the pump failed, or if in tropical conditions, there may not be enough residual pressure for all these operations.

(b) There is no indication of pump failure while the engine is running; accumulator pressure may be used without the pilot being aware of it. If in doubt, do not use the airbrakes and select undercarriage down before flap, to avoid loss of pressure through the flap relief valve.

(c) Unless Mod 1225 or Mod 05/NEAF are embodied, there is no indication of generator failure. If one engine is shut down and the other generator fails, the first indication of failure is a total loss of electrics. For this reason, an engine should not be shut down except in emergency or for practice purposes. In the latter case, practice should be limited to 5 mins duration, in visual contact with a suitable airfield.

74. Stopping an engine

To stop an engine in flight set:—

Throttle	Closed
H.P. cock	OFF
L.P. pump circuit-breaker	Out
Balance cock	As required (see para. 47 (d) (ii))
L.P. cock	Leave on

75. Single-engined flying

(a) Single-engine performance is very good and, after the rudder force has been trimmed out, the aircraft handles easily. The maximum speed at sea level with ventral drop tank fitted is approximately 330 knots. When using full power, the rudder force can be trimmed out at 270 to 280 knots and above. The minimum speed at which the aircraft can be kept straight at sea level with wings level, using full power, is 130 knots; control can be maintained down to 125 knots if 5°–10° of bank is applied towards the live engine. Small angles of bank may be used at any speed below 270 knots, to relieve foot loads.

(b) *Range*

Below 20,000 feet, range may be increased by flying on one engine and cruising at a speed some 30 knots lower than the speed for maximum range on two engines at that altitude.

76. Restarting an engine in flight

(a) *Derwent Mk. 8*

   (i) Attempting to relight an engine at heights above 15,000 ft. is not recommended. Although a successful relight may be achieved, if it is attempted immediately flame extinction occurs, it may, if unsuccessful, jeopardise the chances of a relight below 15,000 ft.

(ii) Check and set:

Altitude	Below 20,000 ft.
Airspeed	To give 1,000–1,200 r.p.m.
Throttle	One-third open
L.P. pump circuit breaker	Made
L.P. cock	On
H.P. cock	OFF

Press the relight button and, after five seconds, set the H.P. cock on, keeping the relight button pressed. When r.p.m. reach 2,000, release the relight button and close the throttle fully. When the engine is running satisfactorily at normal j.p.t., open up to the desired r.p.m.

(iii) If an engine fails to relight within 30 seconds, close the H.P. cock. The next attempt should be made at a lower altitude and a wider throttle opening. At least one minute should elapse between attempts to relight, to allow the engine to dry out.

(iv) No attempt should be made to relight the engine using the normal starter button.

NOTE 1.—If Derwent Mods. 395, 477, 528 are not embodied, the maximum height for relighting is 15,000 ft.

2.—Booster coil life is prolonged if frequent practice relighting is restricted to below 15,000 ft.

(b) *Derwent Mk. 9*

(i) Derwent Mk. 9 engines may be relit at heights up to 25,000 ft.

(ii) Up to 20,000 ft. there is no practical limit to the airspeed at which relighting may be attempted.

(iii) Between 20,000 ft. and 25,000 ft., relighting is more likely to be successful if the airspeed does not exceed 230 knots.

(iv) Unsuccessful attempts to relight at high altitudes should not jeopardise the chances of a successful relight at lower altitudes.

(v) Check and set:—

Throttle	Closed
L.P. pump circuit breaker	Made
L.P. cock	On
H.P. cock	OFF
R.p.m.	Steady

Page 56
Paras.
76
(*contd.*),
77
A.L.4.

Open the H.P. cock, press and hold in the relight button until the r.p.m. rise. Normally light-up occurs in 10–15 seconds but the button may be depressed for a maximum period of 30 seconds. When the engine is running normally, open up to the desired r.p.m.

(vi) If the engine fails to relight, one minute should be allowed to elapse before a further attempt is made at a lower altitude and/or lower airspeed.

77. Asymmetric landing and overshoot

(a) A single-engined landing presents little difficulty. Make a normal circuit, lowering the undercarriage and one-quarter flap in the normal position. Maintain 150 knots across wind until on the final approach, with approximately 13,000 r.p.m. set. Do not lower full flap until a decision to land has been made. Reduce speed on the final approach and use the normal threshold speeds.

(b) *Single-engined overshoot*

  (i) 1. *Starboard engine failed.* Leave the undercarriage down but raise the flaps. Climb away initially at 165 knots.

      2. *Port engine failed.* Raise both the undercarriage and flaps. Climb away initially at 180 knots.

  (ii) Lower the nose to gain speed, at the same time increasing power. Going round again from 150 knots in the above configurations involves no loss of height. The footloads in the overshoot are very heavy.

TARGET TOWING

78. Dart target (Naval aircraft only)

(a) *Take-off*

Line up on the runway centre line and select ¼ flap down. Increase power until the brakes no longer hold, then release the brakes and open up to full power. Raise the nose at 80–85 knots, unstick at 90–100 knots and select undercarriage up. Hold the aircraft down until 130–140 knots is attained, then firmly ease the aircraft into a steep climb, allowing the airspeed to fall to 115–120 knots. At

400 ft. select flap up. At 2,000 ft. ease the nose gently forward to increase speed; if the level-out is made too briskly the target may hit the ground.

NOTE.—The climb to 2,000 ft. is made below safety and critical speeds. Should engine failure occur, the target must be jettisoned.

(b) *Climb*

Climb at 200 knots using 14,100 r.p.m.

(c) *General flying*

In normal circumstances the Dart can hardly be felt by the pilot. Speed must not exceed 200 knots before reaching the operating area and angles of bank used should not exceed 45°. The maximum permissible speed is 275 knots.

(d) *Descending*

Descend at 200 knots, airbrakes out and 12,000 r.p.m. These settings will give a rate of descent of 4,000 ft./min.

(e) *Dropping*

If the Dart is to be dropped in a limited area, it is essential to have R/T guidance, either from a ground controller or a formating aircraft. The lag between manœuvres of the aircraft and response of the target is considerable and it is inadvisable to descend below 2,000 ft. without information of target behaviour. To release the target, switch on the camera master switch and press the camera button on the control column. Release is straightforward and should be made at a speed of 130–150 knots at a tug height of not less than 1,000 ft.

NOTE.—The vertical separation between tug and target with 4,000 ft. of towing cable varies from 800–900 feet at 130 knots to 400 ft. at 300 knots. These figures refer to steady flight conditions and may vary considerably during speed changes.

(f) If the normal release mechanism fails to operate, the target can be released by jettisoning the ventral tank.

(g) *Loss of target*

If the target is lost, the tug speed should be reduced to 200 knots, until the cable has been released.

79. 30 ft. banner targets

(a) Line up on the runway centre-line and set ⅓ flap down. Open up to 11,000 r.p.m. against the brakes, then release the brakes and open up to full power. The nosewheel should be lifted at 95 knots and, once airborne, the aircraft should be climbed initially at 125 knots. When the target is airborne, climb at 200 knots. A speed of 240 knots should not be exceeded and turns should be restricted to below Rate 2. A speed of 200 knots at 8,500 r.p.m., with airbrakes OUT, is recommended for the descent. The target should be released by operating the camera master switch and pressing the camera button on the control column. A speed of 120 knots at 500 feet is recommended for the release.

(b) If the normal release mechanism fails to operate, the target can be released by jettisoning the ventral tank.

80. Winged targets

The take-off, climb and tow are similar to those for the 30 ft. banner target. The target should be landed with the aid of a ground talk-down. A gentle descent with ¾ flap is recommended, aiming to cross the airfield boundary at 450–500 ft., at a speed of 110–115 knots.

PART IV

EMERGENCY HANDLING

81. Engine failure during take-off

(a) The safety speed is 160 knots (165 knots on aircraft fitted with large intakes).

(b) *Engine failure below 130 knots*

When an engine cuts, the aircraft yaws violently and rolls rapidly towards the dead engine. Close the throttles and use coarse rudder to correct the yaw and aileron to regain level flight. When control has been regained, jettison the drop tanks, if fitted. It is recommended that the aircraft be landed straight ahead.

(c) *Engine failure between 130 and 160 knots*

If an engine fails between 130 and 160 knots, reduce the throttle setting and use coarse rudder and aileron to regain level flight. Once level flight has been regained, open up the live engine to full power. The aircraft will climb slowly at 135 knots using full power. The rudder forces are extremely heavy and the use of 5°–10° of bank will help lighten the foot loads; the use of greater angles of bank will adversely affect the climb. Allow the speed to increase to 200 knots and climb.

(d) *Engine failure above safety speed*

Use rudder and aileron as required to maintain level flight. Allow the speed to increase to 200 knots and climb.

82. Engine failure in flight

(a) If an engine fails in flight because of an obvious mechanical defect, carry out the following drill:

H.P. cock	OFF
L.P. cock	Off
L.P. pump	Off
Balance cock	As required

(b) When engine failure has occurred due to flame extinction, do not turn off the L.P. cock, as this may cause damage to the fuel pump and B.P.C.

83. Action in the event of engine fire

If an engine fire warning light comes on, close the throttle immediately, then set:—

H.P. cock	OFF
L.P. cock	Off
L.P. pump	Off

Reduce speed to a practical minimum. Then press the extinguisher button. Should the fire persist, abandon the aircraft. *The engine must not be restarted,* owing to the risk of a further fire with the extinguishers exhausted.

84. Emergency operation of undercarriage, flaps and airbrakes

(a) *Hydraulic failure*

If the starboard engine or the hydraulic pump fails and the hydraulic circuit has not been used since failure, the hydraulic accumulator should have sufficient residual pressure to lower both the undercarriage and flaps when these services are selected. If there is insufficient pressure, due to the prior use of the airbrakes, any of the hydraulic services may be operated by using the handpump. Little resistance is felt when the handpump is used, until the selected service is fully extended. The nosewheel will not lower fully at speeds of 160 knots and above.

(b) *Undercarriage selector lever jamming*

If the undercarriage selector lever jams, it may be due to hydraulic back pressure sticking the rotary valve. Relief is provided by exhausting the hydraulic system. The starboard engine should be throttled back, speed reduced and the airbrakes or flaps operated until the selector is free. If this proves ineffective, flame out the starboard engine and operate the airbrakes.

(c) *Ground-locking with undercarriage up*

If, after selecting DOWN, the port undercarriage is slow to lower, the ground-lock may come into action on the compressed oleo leg and the selector lever will be locked down with the port leg locked up. It will be necessary to operate the ground-lock override before the lever can be freed. The lever should then be selected DOWN in the usual way.

85. Action in the event of electrical failure

(a) *Single generator failure*: If either generator fails, the output of the other is sufficient to keep the battery charged (but see Para 73A(c)).

(b) *Double generator failure*: In the unlikely event of both generators failing, all electrical services will be supplied by the batteries. Switch off all non-essential electrical services, including the L.P. pumps.

86. Hood jettisoning

(a) Reduce speed as much as possible. Then check:—

Seat	Fully lowered
Gunsight	Lowered
Harness	Tight and locked
Helmet	Strap fastened and tight

Before jettisoning the hood assume a position with head well forward and down.

> NOTE.—Unless the aircraft is being flown solo, reduce speed to 150 knots before jettisoning the hood. At speeds above 200 knots it is very difficult, if not impossible, to fly the aircraft from the rear seat with the hood jettisoned.

(b)   (i) In the air, the hood is jettisoned by pulling the jettison handle in either cockpit to its full extent; the handle must be held fully out until the hood has cleared the aircraft.

(ii) On the ground, the normal opening levers should be used, so that the hood can be swung over the side.

(iii) In the air or on the ground, if the above methods fail, pull the jettison handle to free the side toggles, then release it to disengage the rear hooks. The hood can then be pushed off to starboard to clear the assister arm. If the rear hooks fail to disengage, they can be operated by pulling one of the toggles on the starboard side of each cockpit or, on the ground, by the external release on the port side of the fuselage.

(c) If time permits, the hood should be jettisoned before a crash landing or ditching.

87. Action in the event of hood opening in flight

If the hood opens inadvertently in flight, do not take jettison action, as the control surfaces may be damaged. Correct the strong yaw to starboard by rudder and differential use of the engines. Reduce speed to at least 200 knots as quickly as possible. Below this speed, adequate control can be maintained and a normal landing made.

88. Drop-tank jettisoning

(a) The ventral drop tank may be jettisoned at any speed up to 345 knots by pulling the T-handle at the left-hand side of the instrument panel in the front cockpit.

(b) The wing-drop tanks may be jettisoned at any speed up to 260 knots by pulling back the lever at the starboard side of the seat in the front cockpit. When jettisoning the wing drop tanks the control column should be held firmly at the moment of release, as the stick force required to maintain lateral control is heavy if one tank fails to jettison. In this event, speed should be reduced as much as practicable to reduce the aileron stick force.

89. Failure of wing-drop tank to jettison

(a) If a wing-drop tank fails to jettison, ample aileron control will be available, even at the stalling speed. The force necessary to maintain the wings level increases with speed and is heavy above 260 knots. This applies whether the tank is full or empty and whether flying on one or two engines.

(b) Make a normal circuit and landing. If the tank is full, the landing run will be longer, since a degree of differential braking will be necessary to keep the aircraft straight.

90. Forced landing without power

(a) Experience suggests that it is preferable to lower the undercarriage when making a forced landing on an aerodrome or in open country. In the down position, it absorbs much if not all of the initial impact, assists in retarding the aircraft and it may be retracted after touchdown if necessary.

(b) The best gliding speed is 170 knots. The rate of descent with undercarriage down, flaps up, at this speed, is approximately 2,500 ft./min.

(c) If a forced landing on an airfield is being made, aim to be at 2,000 ft. at the end of the downwind leg opposite the caravan.

(d) Maintain 150 knots crosswind; lower full flap when required (be prepared to pump the flaps down). Cross the threshold at 120 knots.

63 (e) Carry out the following actions during the circuit:—

Airbrakes	IN
Undercarriage	Down below 175 knots. Lever fully down. Select ground-lock override up.
Fuel	H.P. and L.P. cocks OFF.
Flaps	As required.
Harness	Seat harness tight and locked. Parachute harness and dinghy released.
Undercarriage indicator	Three green lights. Nosewheel indicator showing.
Brakes	Pressure.
Hood	Jettison (at pilot's discretion).

91. Ditching

Tests indicate that in calm weather the aircraft should ditch well.

(a) Jettison the wing and ventral tanks.

(b) Jettison the hood.

(c) Tighten the seat harness, fasten and tighten the helmet strap and release the parachute harness.

(d) Make a normal powered approach with the undercarriage up and full flap selected. Flatten out the approach just before touchdown to keep the rate of descent to a minimum. Ditch along the swell or, if the swell is not steep, into wind.

92. **Controlled descent through cloud with an unserviceable A.S.I.**

(a) A guide to making a safe descent without an A.S.I. is given below:—

Condition of flight	Settings	Speed attained (knots)
Fast rate of descent	11,000 r.p.m. Airbrakes OUT 3,500 ft./min. on V.S.I.	230 ± 10
Slow rate of descent	11,000 r.p.m. Airbrakes OUT ⅓ flap 1,000 ft./min. on V.S.I.	170 ± 10
Level flight at 1,500 ft.	11,000 r.p.m. Airbrakes IN ⅓ flap	170 ± 10
Final descent	11,000 r.p.m. Undercarriage down ⅔ flap 700 ft./min. on V.S.I.	135 ± 10

93. **Abandoning the aircraft in flight**

(a) Reduce speed as much as possible, then check:—

Throttle(s)	Closed
Seat	Fully lowered
Gunsight	Lowered
Harness	Tight and locked
Helmet	Strap fastened and tight
Hood	Jettison

Then:

Lead	Disconnected and stowed
Harness	Released
Airbrakes	IN (this is essential)

(b) (i) Abandon the aircraft by diving over the inboard trailing edge of the wing.

(ii) Do not attempt to abandon the aircraft by dropping out from the inverted position.

(iii) If the aircraft is spinning, abandon the aircraft by diving over the side away from the axis of spin.

(c) The hood opening strut remains on the aircraft after the hood has been jettisoned and forms a considerable obstruction on the starboard side for the occupant of the front seat. Therefore, unless the aircraft is in a spin to the left, the occupant of the front seat should leave the aircraft on the port side.

PART V

OPERATING DATA

94. Weight and C.G. data

(a) The weight and C.G. limitations are given in Part II, para. 46 (c) (d).

(b) The aircraft should always be loaded in accordance with the instructions given in A.P.2210G, Vol. 1, Section 4, Chapter 1.

Page 66
94 (c)
A.L.5. (c) The aircraft is at its maximum all-up weight with two pilots, full internal fuel, full ventral and wing drop tanks. The aircraft is at maximum landing weight with two pilots, full internal fuel, empty ventral and wing drop tanks.

(d) C.G. movement with consumption of fuel is as follows:—

Wing drop tank fuel	..	..	Slightly forward
Ventral tank fuel ..	..	..	Aft
Internal fuel	..	..	Forward

95. Pressure error corrections

The following corrections (in knots) should be applied to the A.S.I. reading to obtain R.A.S.

From To	130 170	170 220	220 300	300 510
Subtract ..	1	0	—	—
Add	—	0	2	4

96. Fuel consumptions

The following are the approximate fuel consumptions in gallons/hour for various altitudes and power settings. The figures quoted assume the use of AVTAG (7.7 lb./gall.); if AVTUR is used, the figures are reduced by 4 per cent.

Configuration	Height ft.	At max. r.p.m.	At max. cont. r.p.m. 14,100	At best range speed	At best endurance speed
Clean	S.L.	1,240	1,009	383	285
	10,000	950	774	322	250
	20,000	750	624	269	215
	30,000	560	480	237	200
	35,000	480	401	228	197
Ventral tank	S.L.	1,240	1,009	405	300
	10,000	950	774	343	260
	20,000	750	624	286	225
	30,000	560	480	254	212
	35,000	480	401	239	203
Ventral and Tip tanks	S.L.	1,230	1,009	419	320
	10,000	940	774	349	270
	20,000	740	624	299	250
	30,000	550	480	270	230
	35,000	470	401	258	220

97. Take-off distances

The approximate distances (in yards) to unstick and to clear 50 ft., are given below:—

(a) *Clean aircraft*

Altitude feet	Windspeed knots		Temperature °C.			
			0	15	30	45
0	Zero	Ground run	370	455	540	625
		To 50 feet	740	905	1,067	1,250
	30	Ground run	190	240	280	325
		To 50 feet	545	670	800	930
1,000	Zero	Ground run	395	490	570	665
		To 50 feet	770	960	1,135	1,320
	30	Ground run	205	250	295	345
		To 50 feet	575	720	855	985
2,000	Zero	Ground run	420	515	605	705
		To 50 feet	815	1,015	1,200	1,405
	30	Ground run	220	270	315	365
		To 50 feet	610	760	910	1,045

PART V—OPERATING DATA

(b) *With ventral tank*

Altitude feet	Windspeed knots		Temperature °C.			
			0	15	30	45
0	Zero	Ground run	480	585	700	835
		To 50 feet	935	1,150	1,375	1,640
	30	Ground run	260	320	385	460
		To 50 feet	710	865	1,040	1,245
1,000	Zero	Ground run	510	620	745	890
		To 50 feet	980	1,210	1,465	1,745
	30	Ground run	285	345	410	490
		To 50 feet	755	930	1,115	1,325
2,000	Zero	Ground run	535	660	795	945
		To 50 feet	1,030	1,285	1,555	1,855
	30	Ground run	305	360	430	520
		To 50 feet	795	980	1,180	1,410

(c) *With ventral tank and wing drop tanks*

Altitude feet	Windspeed knots		Temperature °C.			
			0	15	30	45
0	Zero	Ground run	625	760	910	1,095
		To 50 feet	1,210	1,495	1,800	2,135
	30	Ground run	355	430	520	625
		To 50 feet	930	1,160	1,390	1,665
1,000	Zero	Ground run	655	810	970	1,165
		To 50 feet	1,285	1,590	1,915	2,275
	30	Ground run	380	460	550	665
		To 50 feet	980	1,230	1,475	1,765
2,000	Zero	Ground run	690	855	1,030	1,240
		To 50 feet	1,360	1,680	2,035	2,420
	30	Ground run	405	490	590	705
		To 50 feet	1,035	1,295	1,575	1,865

98. Flight planning data

(a) *Flight planning tables*

The tables show the flight planning data for the aircraft in its clean condition, with ventral tank, and with ventral and wing drop tanks. The following information is given:—

(i) *Climbing*

The climb tables give data for climbs in I.S.A. conditions, using the speeds given in para. 57. Since climb performance is dependent on temperature, corrections are given for each 10°C. rise in temperature above I.S.A.

(ii) *Cruising at best range speed*

Each separate block in the cruise tables shows:—

1. The speed for maximum range, the approximate A.N.M.P.G. and the approximate fuel consumption, in gallons per minute, for the particular height. In addition, a speed band is given; use of any speed within this range should not cause more than a 5 per cent reduction in range.

2. The range obtainable for various amounts of available fuel when flying at best range speed for that height. The range given is to the point of letdown, allowance being made for the descent fuel required.

3. The range obtainable for various amounts of available fuel, including the distance covered on the climb, if a climb is made to another altitude. In this case the climb must be made at the speeds given in para. 57 and the flight continued at the new altitude, at the best speed for that height.

NOTE.—The range at any altitude is independent of temperature but is dependent on the *weight* of fuel carried.

(iii) The descent table gives the data for descending from one height to another.

(b) *Use of the tables*

(i) *Pre-flight planning*

Enter the cruise data table in the sea level block, at the fuel state applying immediately after take-off.

Select the height at which maximum range is available at that fuel state. The distance available includes distance covered on the climb but not on the descent. (Absolute maximum range is obtained by adding on the descent distance, provided that the let-down starts at that distance from the destination.) For short-range flights, inspect the sea level block and select the height at which the distance to be covered requires the least amount of fuel. This is the best altitude for the flight.

(ii) *In-flight planning*

At any stage of the flight, the available range may be ascertained by applying the fuel state to the level flight range in the particular block. If an increase in range is required, or if a climb has to be made, the new available range may be obtained by entering the existing altitude block at the appropriate fuel state and moving vertically downwards within the block until the new altitude is reached. Figures in heavy type indicate the best altitude for the maximum increase in range. Above these heights, no further increase in range is possible. If a descent is necessitated, the new range is shown by moving direct from the existing altitude level flight range for the appropriate fuel state to the new altitude level flight range.

(c) *Cruise data charts*

The cruise data charts show aircraft performance and fuel consumption in level flight at various heights and r.p.m.

99. Endurance

At any altitude, maximum endurance will be obtained by flying at the I.A.S. which requires the lowest r.p.m. to maintain height. Increase in altitude gives increase in endurance; however, the overall endurance may not be improved by climbing above 20,000 ft., except when maximum fuel is available, owing to the proportion of fuel used on the climb and the descent. All climbs should be made at full power (within the jet pipe limitations) and the descent made as quickly as practicable.

CLIMB DATA—CLEAN AIRCRAFT

Climb at full throttle throughout, at speeds quoted in para. 57

From	To	Time Mins.	Fuel Galls. (AVTAG)	Distance N.M.
Sea Level	10,000 ft.	1½	22	7
	20,000 ft.	3	44	18
	30,000 ft.	5½	65	33
	35,000 ft.	7	79	43
10,000 ft.	20,000 ft.	1½	22	11
	30,000 ft.	4	43	26
	35,000 ft.	5½	57	36
20,000 ft.	30,000 ft.	2½	21	15
	35,000 ft.	4	35	25
30,000 ft.	35,000 ft.	1½	14	10

Temperature correction. For each 10° C. rise in ambient temperature, add the following:—

From	To	Time	Fuel	Distance
Sea Level	10,000 ft.	0.1	1.3	0.6
10,000 ft.	20,000 ft.	0.2	2.8	1.9
20,000 ft.	30,000 ft.	0.5	5.4	3.7
30,000 ft.	35,000 ft.	0.7	6.9	5.0

FUEL CONTENTS   . .   . .   325 gallons

TAXY AND TAKE-OFF ALLOWANCE   . .   . .   40 gallons

LANDING ALLOWANCE (excluding descent fuel)   . .   60 gallons

DESCENT DATA—CLEAN AIRCRAFT

Airbrakes IN, throttles closed, 0.6M

From	To	Time Mins.	Fuel Galls. (AVTAG)	Distance N.M.
35,000 ft.	30,000 ft.	$1\frac{1}{2}$	2	11
	20,000 ft.	4	6	26
	10,000 ft.	$5\frac{1}{2}$	10	38
	Sea Level	7	16	46
30,000 ft.	20,000 ft.	$2\frac{1}{2}$	4	15
	10,000 ft.	4	8	27
	Sea Level	$5\frac{1}{2}$	14	35
20,000 ft.	10,000 ft.	$1\frac{1}{2}$	4	12
	Sea Level	3	10	20
10,000 ft.	Sea Level	$1\frac{1}{2}$	6	8

CRUISE DATA—CLEAN AIRCRAFT

FUEL AVAILABLE Galls. (AVTAG)	285	250	200	150	100
Sea Level — Range	145	125	90(22)	60(14)	**25(6)**
10,000 ft.	180	150	105	60	20
20,000 ft.	220	180	120	**60**	—
30,000 ft.	260	205	**125**	50	—
35,000 ft.	**265**	**205**	120	—	—

Sea Level
ANM/Gallon—0.655
Gall./min.—6.4
Best Range I.A.S.—275 Kts.
95% Range I.A.S.—180 Kts.–370 Kts.

	285	250	200	150	100
10,000 ft. — Range	—	160	120(25)	75(16)	**30(6)**
20,000 ft.	—	200	140	**80**	20
30,000 ft.	—	230	**155**	75	—
35,000 ft.	—	**235**	150	65	—

10,000 ft.
ANM/Gallon—0.88
Gall./min.—5.4
Best Range I.A.S.—263 Kts.
95% Range I.A.S.—180 Kts.–340 Kts.

	285	250	200	150	100
20,000 ft. — Range	—	—	155(29)	95(18)	**35(7)**
30,000 ft.	—	—	175	**100**	25
35,000 ft.	—	—	**175**	90	—

20,000 ft.
ANM/Gallon—1.19
Gall./min.—4.5
Best Range I.A.S.—235 Kts.
95% Range I.A.S.—180 Kts.–300 Kts.

	285	250	200	150	100
30,000 ft. — Range	—	—	195(32)	**115(19)**	**40(7)**
35,000 ft.	—	—	**200**	112	25

30,000 ft.
ANM/Gallon—1.54
Gall./min.—3.95
Best Range I.A.S.—230 Kts. (0.62M)
95% Range I.A.S.—180 Kts.–270 Kts. (0.49M–0.72M)

	285	250	200	150	100
35,000 ft. — Range	—	—	**210(30)**	**125(18)**	**40(6)**

35,000 ft.
ANM/Gallon—1.71
Gall./min.—3.8
Best Range I.A.S.—220 Kts. (0.66M)
95% Range I.A.S.—180 Kts.–253 Kts. (0.55M–0.75M)

FUEL AVAILABLE Galls. (AVTUR)	274	241	193	144	96

Figures in brackets denote time remaining for cruise (minutes)

CLIMB DATA—VENTRAL TANK

Climb at full throttle throughout, at speeds quoted in para. 57

From	To	Time Mins.	Fuel Galls. (AVTAG)	Distance N.M.
Sea Level	10,000 ft.	1½	24	10
	20,000 ft.	3½	49	24
	30,000 ft.	6½	76	44
	35,000 ft.	8½	91	55
10,000 ft.	20,000 ft.	2	25	14
	30,000 ft.	5	52	34
	35,000 ft.	7	67	45
20,000 ft.	30,000 ft.	3	27	20
	35,000 ft.	5	42	31
30,000 ft.	35,000 ft.	2	15	11

Temperature correction. For each 10 C. rise in ambient temperature, add the following:—

From	To	Time	Fuel	Distance
Sea Level	10,000 ft.	0.1	1.5	0.9
10,000 ft.	20,000 ft.	0.3	3.7	2.5
20,000 ft.	30,000 ft.	0.7	7.0	5.8
30,000 ft.	35,000 ft.	1.1	9.1	8.0

FUEL CONTENTS .. .. 500 gallons

TAXY AND TAKE-OFF ALLOWANCE .. .. 40 gallons

LANDING ALLOWANCE (excluding descent fuel) .. 60 gallons

DESCENT DATA
As for clean aircraft

CRUISE DATA—VENTRAL TANK

FUEL AVAILABLE Galls. (AVTAG)		465	400	300	250	200	150	100
Sea Level	Range	260	220	155	125	90 (21)	60 (14)	**25** (6)
ANM/Gallon—0.645	10,000 ft.	335	280	195	150	105	65	20
Gall./min.—6.75	20,000 ft.	440	365	250	190	**135**	**75**	—
Best Range I.A.S.—259 Kts.	30,000 ft.	515	420	270	200	125	50	—
95% Range I.A.S.—190 Kts.-345 Kts.	35,000 ft.	**555**	**445**	**280**	**200**	120	—	—
10,000 ft.	Range	—	290	205	160	115 (23)	75 (15)	**39** (6)
ANM/Gallon—0.87	20,000 ft.	—	370	255	195	140	**80**	20
Gall./min.—5.75	30,000 ft.	—	445	295	220	**150**	75	—
Best Range I.A.S.—256 Kts.	35,000 ft.	—	**475**	**310**	**225**	145	45	—
95% Range I.A.S.—190 Kts.-337 Kts.								
20,000 ft.	Range	—	380	265	210	150 (27)	95 (17)	**35** (6)
ANM/Gallon—1.16	30,000 ft.	—	465	320	245	170	**95**	20
Gall./min.—4.8	35,000 ft.	—	**500**	**335**	**255**	**170**	85	—
Best Range I.A.S.—245 Kts.								
95% Range I.A.S.—190 Kts.-300 Kts.								
30,000 ft.	Range	—	—	335	260	**190** (30)	**115** (18)	**40** (6)
ANM/Gallon—1.48	35,000 ft.	—	—	360	**275**	175	110	30
Gall./min.—4.2								
Best Range I.A.S.—230 Kts. (0.63M)								
95% Range I.A.S.—190 Kts.-273 Kts. (.52M-.72M)								
35,000 ft.	Range	—	—	370	290	**205** (31)	**125** (19)	**40** (6)
ANM/Gallon—1.66								
Gall./min.—4.0								
Best Range I.A.S.—223 Kts. (0.68M)								
95% Range I.A.S.—185 Kts.-255 Kts. (.56M-.75M)								
FUEL AVAILABLE Galls. (AVTUR)		448	385	289	241	193	144	96

Figures in brackets denote time remaining for cruise (minutes)

CLIMB DATA—VENTRAL AND WING DROP TANKS

Climb at full throttle throughout, at speeds quoted in para. 57

From	To	Time Mins.	Fuel Galls. (AVTAG)	Distance N.M.
Sea Level	10,000 ft.	2	29	12
	20,000 ft.	4½	60	30
	30,000 ft.	8½	95	52
	35,000 ft.	11	115	68
10,000 ft.	20,000 ft.	2½	31	18
	30,000 ft.	6½	66	40
	35,000 ft.	9	86	56
20,000 ft.	30,000 ft.	4	35	22
	35,000 ft.	6½	55	38
30,000 ft.	35,000 ft.	2½	20	16

Temperature correction. For each 10 C. rise in ambient temperature, add the following:—

From	To	Time	Fuel	Distance
Sea Level	10,000 ft.	0.1	1.8	1.2
10,000 ft.	20,000 ft.	0.4	5.0	3.3
20,000 ft.	30,000 ft.	1.1	9.4	8.0
30,000 ft.	35,000 ft.	1.8	13.0	12.0

FUEL CONTENTS .. .. 700 gallons

TAXY AND TAKE-OFF ALLOWANCE .. .. 40 gallons

LANDING ALLOWANCE (excluding descent fuel) .. 60 gallons

DESCENT DATA

As for clean aircraft

CRUISE DATA—VENTRAL AND WING DROP TANKS

FUEL AVAILABLE	Galls. (AVTAG)	665	600	500	400	300	200	100
Sea Level ANM/Gallon—0.62 Gall./min.—7.0 Best Range I.A.S.—256 Kts. 95% Range I.A.S.—200 Kts.–360 Kts.	Range	375	335	270	210	150	85 (20)	**25** (6)
	10,000 ft.	480	425	345	260	180	100	15
	20,000 ft.	592	525	420	315	210	**105**	—
	30,000 ft.	720	635	500	365	230	95	—
	35,000 ft.	**780**	**680**	**530**	**380**	**230**	80	—
10,000 ft. ANM/Gallon—0.82 Gall./min.—5.8 Best Range I.A.S.—255 Kts. 95% Range I.A.S.—180 Kts.–340 Kts.	Range	—	435	355	275	190	110 (23)	**30** (6)
	20,000 ft.	—	540	435	330	225	120	—
	30,000 ft.	—	660	525	390	255	**120**	—
	35,000 ft.	—	**715**	**565**	**415**	**265**	115	—
20,000 ft. ANM/Gallon—1.045 Gall./min.—5.0 Best Range I.A.S.—235 Kts. 95% Range I.A.S.—180 Kts.–305 Kts.	Range	—	555	450	345	240	135 (26)	**30** (6)
	30,000 ft.	—	685	550	415	280	**145**	—
	35,000 ft.	—	**740**	**590**	**440**	**290**	140	—
30,000 ft. ANM/Gallon—1.35 Gall./min.—4.5 Best Range I.A.S.—230 Kts. (0.62M) 95% Range I.A.S.—180 Kts.–270 Kts. (.5M–.72M)	Range	—	—	575	440	305	170 (28)	35 (6)
	35,000 ft.	—	—	**620**	**470**	**320**	**170**	20
35,000 ft. ANM/Gallon—1.54 Gall./min.—4.3 Best Range I.A.S.—215 Kts. (0.65M) 95% Range I.A.S.—190 Kts.–240 Kts. (0.57M–0.72M)	Range	—	—	635	485	335	**185** (29)	35 (5)
FUEL AVAILABLE	Galls. (AVTUR)	640	578	481	385	289	193	96

CLEAN AIRCRAFT

FUEL CONSUMPTION GALL/HR
AT BEST ENDURANCE SPEED (AVTAG)

MEAN CRUISE
WEIGHT ~ 12,950 lb

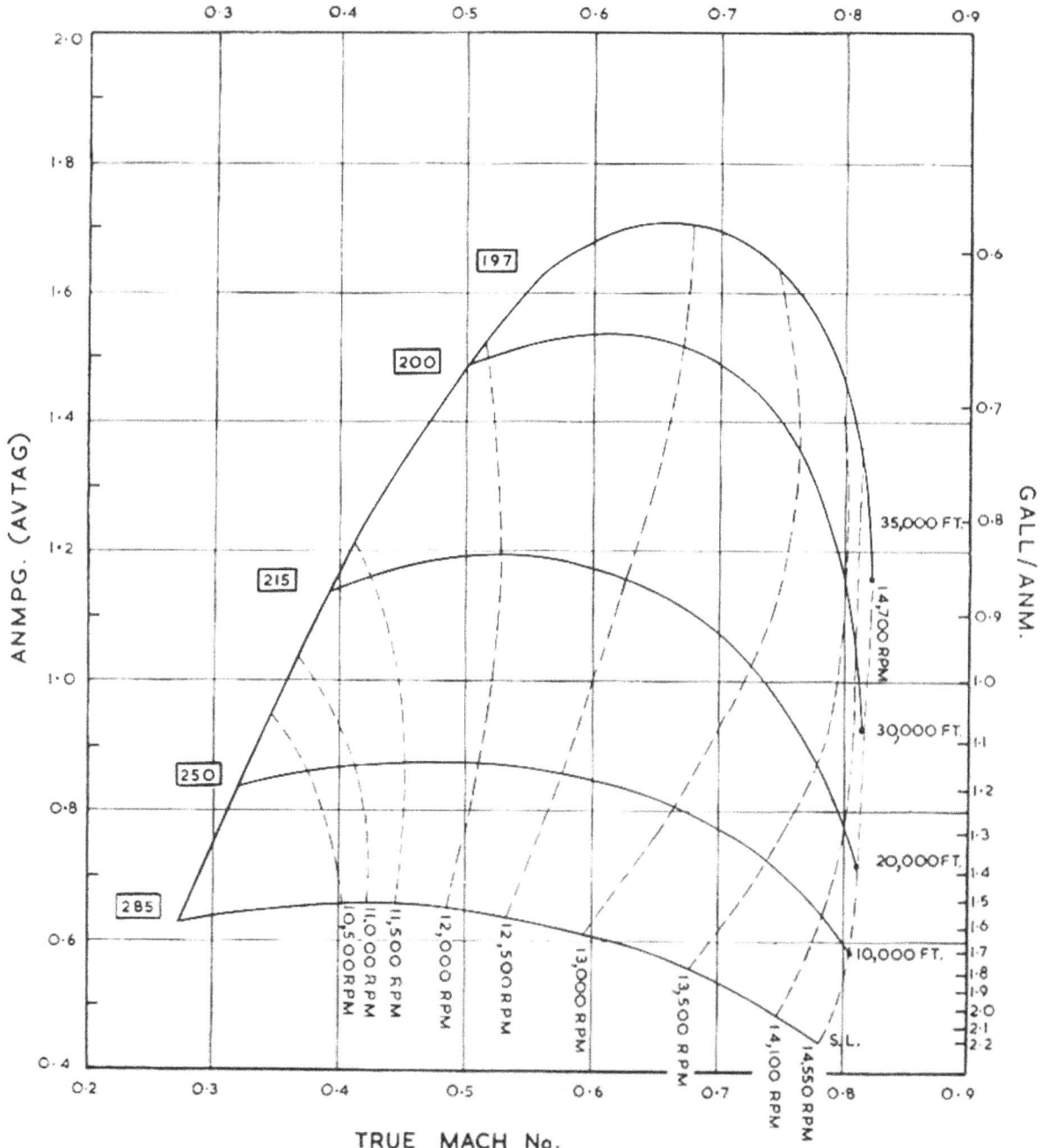

CRUISE DATA CHART

AIRCRAFT   WITH   VENTRAL   TANK

FUEL CONSUMPTION GALL / HR
AT BEST ENDURANCE SPEED (AVTAG)

MEAN CRUISE
WEIGHT ~ 13,900 lb

TRUE MACH. No.

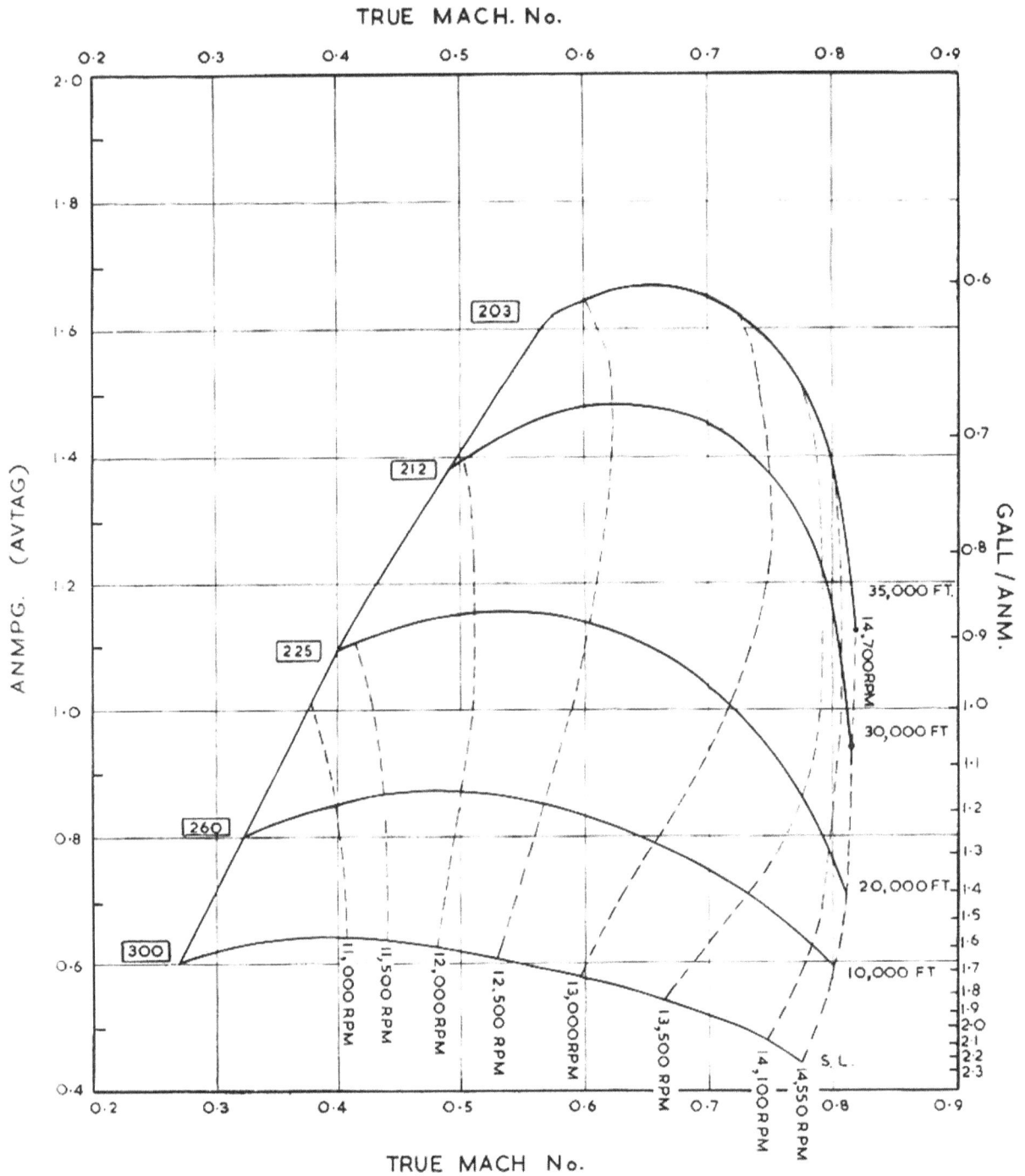

ANMPG. (AVTAG)

GALL / ANM.

203

212

225

260

300

35,000 FT.
14,700 RPM
30,000 FT
20,000 FT
10,000 FT
S.L.

11,000 RPM
11,500 RPM
12,000 RPM
12,500 RPM
13,000 RPM
13,500 RPM
14,100 RPM
14,550 RPM

TRUE MACH No.

CRUISE DATA CHART

79

AIRCRAFT WITH VENTRAL AND WING TANKS

☐ FUEL CONSUMPTION GALL/HR
AT BEST ENDURANCE SPEED (AVTAG)

MEAN CRUISE
WEIGHT ~ 14,850 lb

TRUE MACH No.

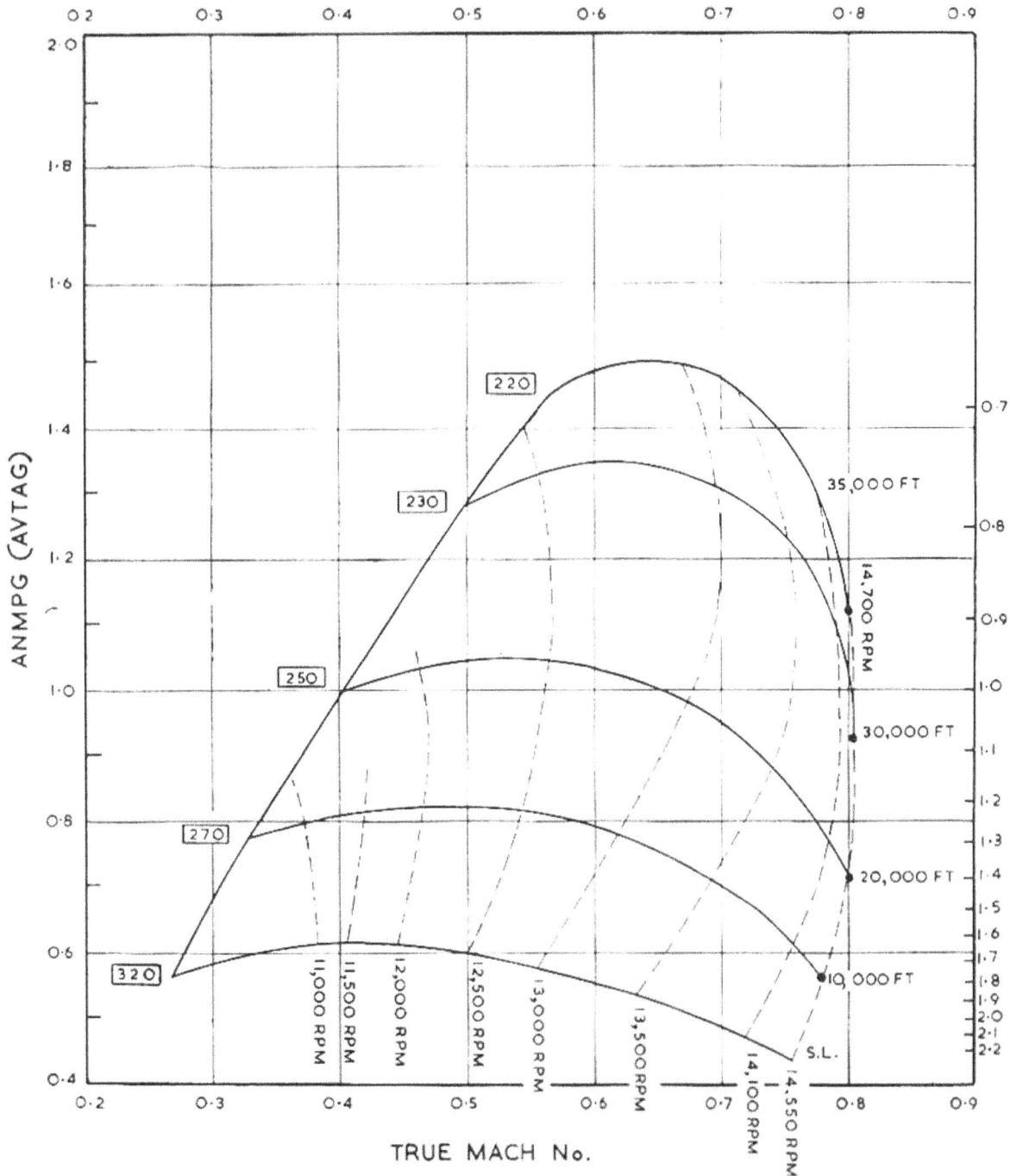

CRUISE DATA CHART

80

PART VI

ILLUSTRATIONS

KEY TO FIGURES 1 to 6

Front cockpit

1. Port engine L.P. cock lever
2. No. 2 V.H.F. controller
3. Port engine H.P. cock lever
4. Port engine relight button
5. Mounting for camera footage indicator
6. Micro-switch for operating undercarriage warning light
7. Throttle micro-switch in starting circuit
8. Pneumatic triple-pressure gauge
9. I.F.F. manual, G-auto and G-D switches
10. V.H.F. change-over switch
11. Hood opening lever
12. Throttle levers (two)
13. Direct vision panel control handwheel
14. Cockpit ventilation control
15. Undercarriage emergency override lever
16. Auxiliary red lamps dimmer switch
17. Airbrakes control lever
18. Landing lamp switch
19. L.P. pumps circuit breakers (two)
20. Engine starter pushbuttons (two)
21. L.P. pumps ammeter test socket
22. Map case
23. No. 1 V.H.F. controller
24. L.P. pumps test pushbuttons (two)
25. Elevator trimming tab handwheel
26. Elevator trimming tab indicator
27. Rudder trimming tab indicator
28. Rudder trimming tab control
29. Fuel balance cock
30. Drop tanks transfer warning light
31. Undercarriage selector lever
32. Flaps selector lever
33. Rudder pedal adjuster release
34. Flaps position indicator
35. Drop tanks control
36. Port engine fire-warning light
37. G.G.S. spare filaments
38. G.G.S. control switch
39. Mk. 4F compass
40. G.G.S. emergency retraction swit
41. G.G.S. master circuit breaker
42. E.2.A compass
43. Starboard engine fire-warning light
44. Hood jettison handle

45. Engine fire-extinguisher pushbuttons (two)
46. Clock
47. Oxygen regulator
48. Press-to-transmit switch
49. Camera control button
50. Instrument panel red lamps dimmer switch
51. Starboard engine oil pressure gauge
52. Fuel contents gauge (rear compartment)
53. A.C. voltmeter
54. Fuel contents gauge (front compartment)
55. Port engine oil pressure gauge
56. Undercarriage warning light
57. Instrument panel u/v lamps dimmer switch
58. Undercarriage position indicator
59. Compass master switch
60. Resin lights switch
61. Pressure-head heater switch
62. Navigation lights switch
63. Camera master switch
64. Compass emergency A.C. switch
65. Compass emergency D.C. switch
66. Identification lights colour selector switch
67. Identification lights pushbutton
68. G.G.S. selector-dimmer control
69. I.F.F. D switch
70. I.F.F. F switch
71. Cockpit heating control
72. I.F.F. controller
73. Hood starboard catches indicator
74. Safety harness release
75. Windscreen heater switch
76. Emergency lamp dimmer switch
77. Hood rear hooks emergency release
78. Starboard engine relight button
79. Starboard engine H.P. cock
80. Starboard engine L.P. cock
81. Mic/tel socket
82. Windscreen de-icing handpump
83. Wing drop tanks jettison lever
84. Oxygen tube
85. Hydraulic emergency handpump
86. Bracket for controls locking rod

87. Port engine L.P. cock lever
88. Port engine relight button
89. Port engine H.P. cock lever
90. Throttle levers (two)
91. Undercarriage emergency override lever
92. Guard for hood opening lever
93. Hood opening lever
94. Jet pipe temperature gauge
95. Airbrakes control lever
96. I.F.P.E. goggles stowage
97. Elevator trimming tab handwheel
98. Elevator trimming tab indicator
99. Rudder trimming tab indicator
100. Rudder trimming tab control
101. Auxiliary red lamps dimmer switch
102. Undercarriage selector lever
103. Flaps selector lever
104. Rudder pedals adjuster release
105. Flaps position indicator
106. Hood unlocked warning light
107. J.p.t. gauge lamp dimmer switch
108. Hood jettison handle
109. Oxygen regulator
110. E.2.A compass
111. Fuel contents gauges (two)
112. Undercarriage warning light
113. Undercarriage position indicator
114. Clock
115. U/v lamps dimmer switch
116. Red lamps dimmer switch
117. Emergency lamps dimmer switch
118. Mic/tel socket
119. V.H.F. mute switch
120. Hood rear hooks emergency release
121. Safety harness release
122. Starboard engine relight button
123. Starboard engine H.P. cock lever
124. Starboard engine L.P. cock lever
125. Map case
126. Mk. 4F compass corrector control box
127. Oxygen tube

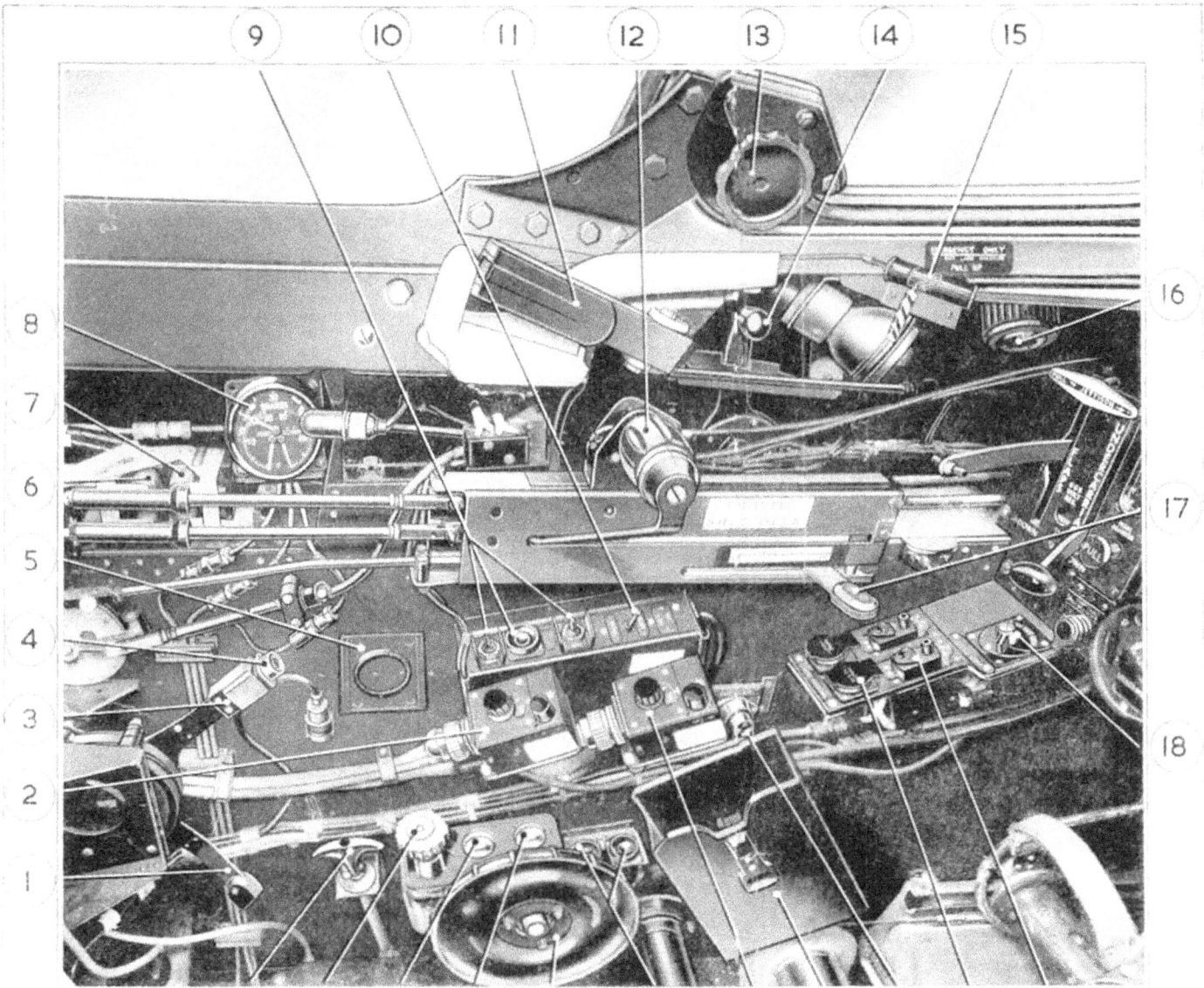

FRONT COCKPIT—PORT SIDE

FIG
1

FIG
1

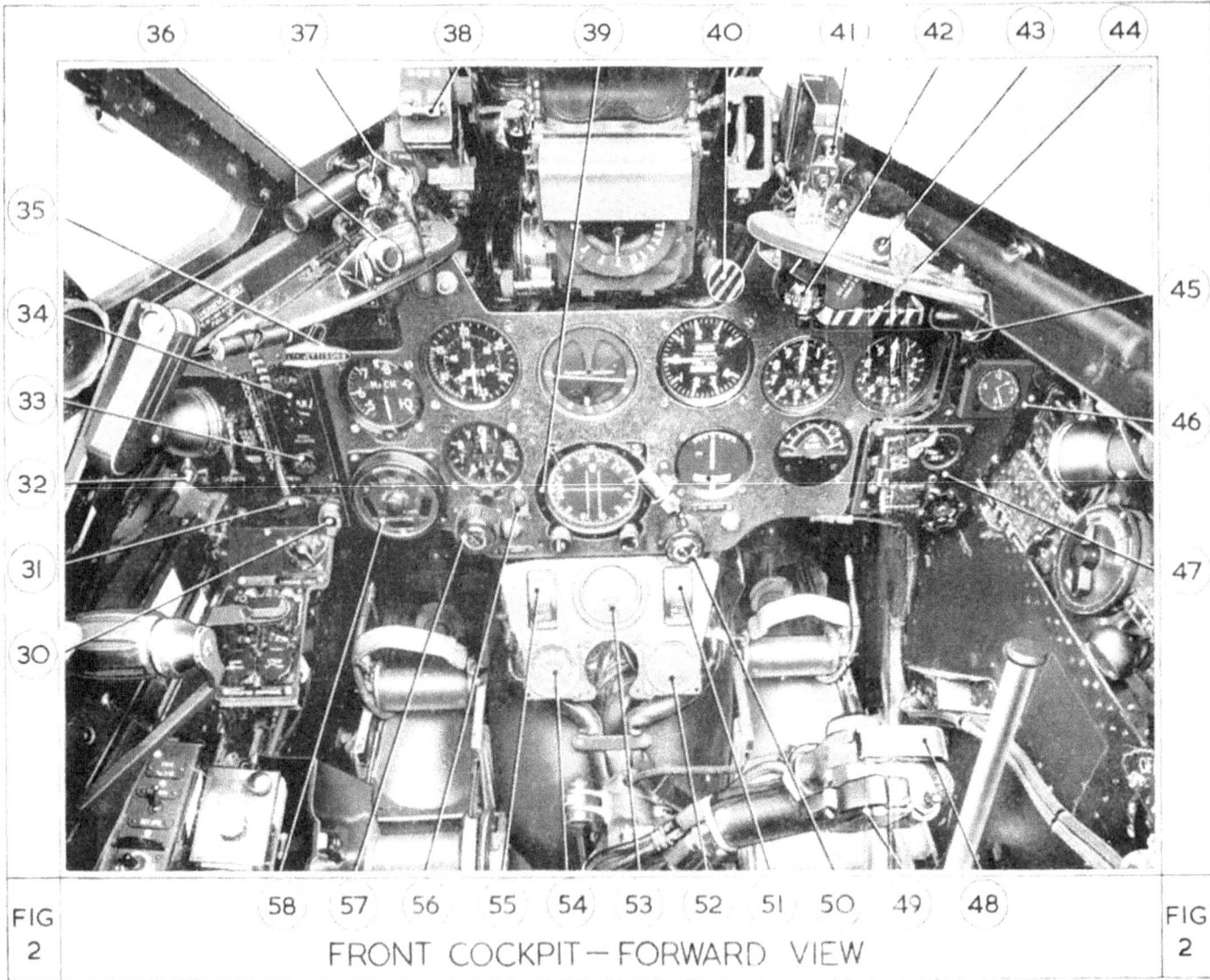

FRONT COCKPIT—FORWARD VIEW

FIG
2

FIG
2

FRONT COCKPIT—STARBOARD SIDE

FIG
3

FIG
3

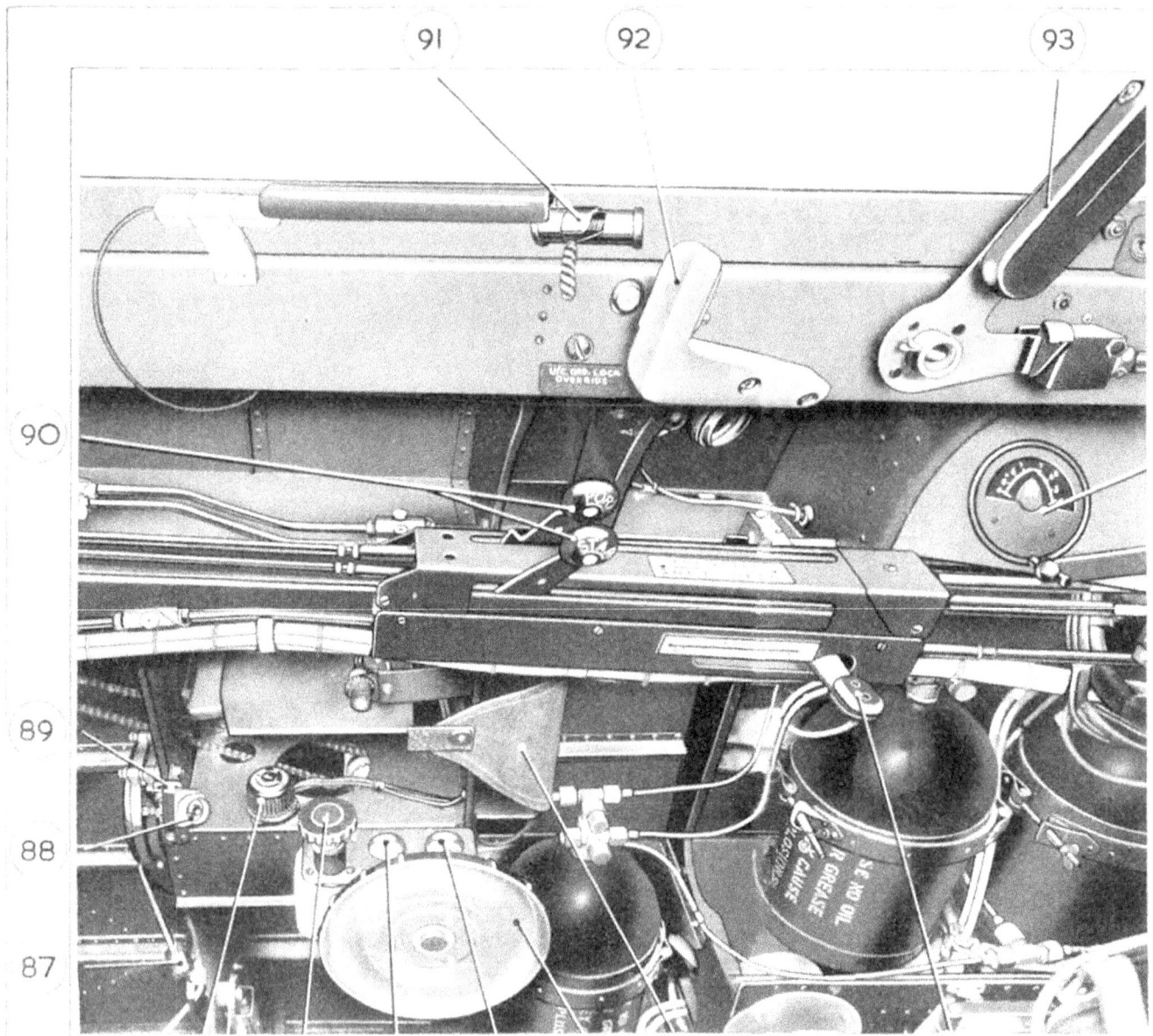

REAR COCKPIT — PORT SIDE

FIG 4

FIG 4

106 107

105

104

103

102

HOOD
UNLOCKED

108

109

110

REAR COCKPIT—FORWARD VIEW

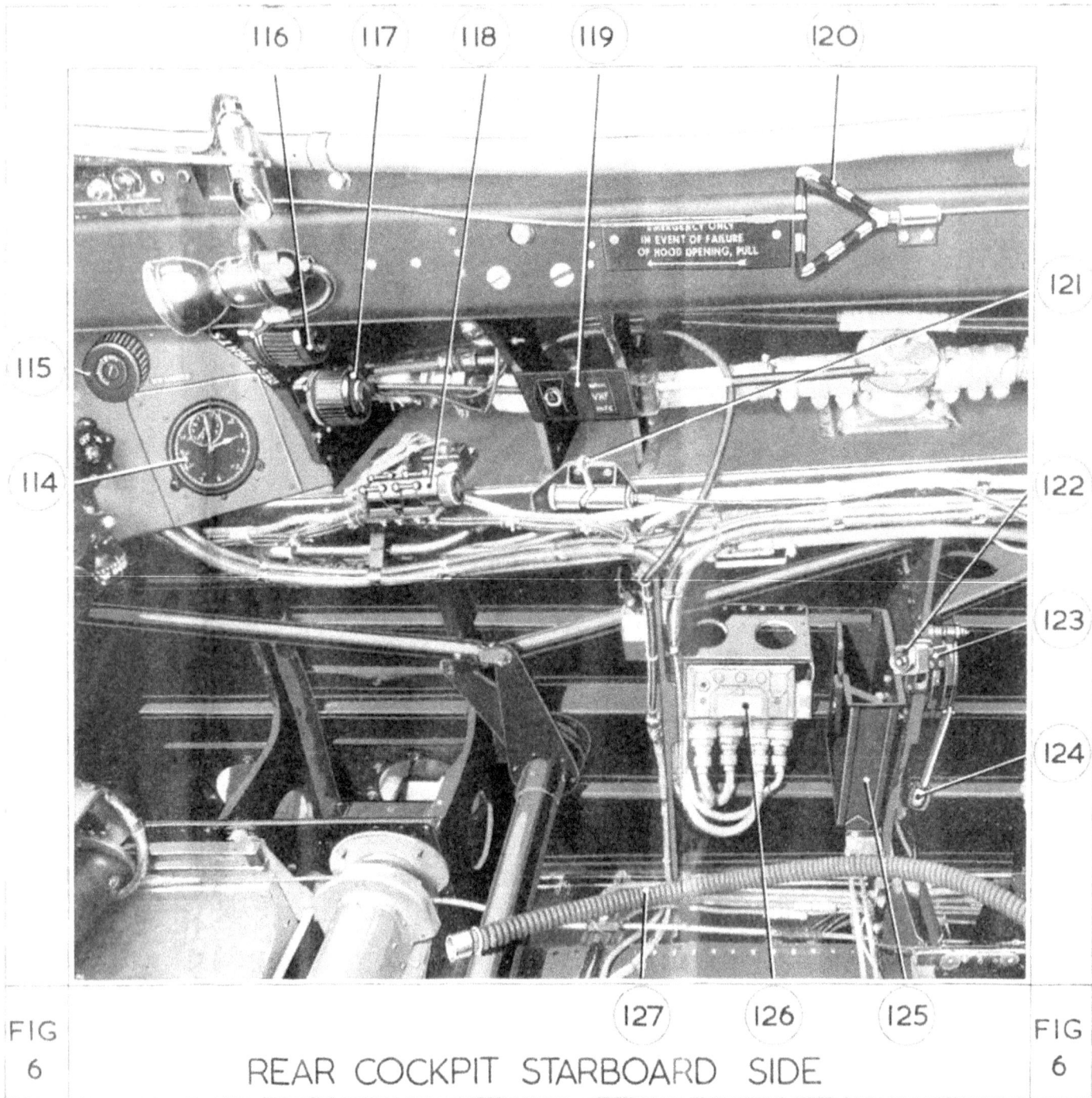

116 117 118 119 120

115

121

114

122

123

124

127 126 125

FIG 6

REAR COCKPIT STARBOARD SIDE

FIG 6

VITAL DRILLS

ENGINE FAILURE IN FLIGHT

(a) Mechanical Failure

H.P. and L.P. cocks	Off
L.P. pump	Off
Balance cock	As required

Do not attempt to relight.

(b) Relighting Derwent Mk. 8 Engines

Altitude	Below 20,000 ft.
Airspeed	To give 1,000–1,200 r.p.m.
Throttle	⅓ open
L.P. pump	On
L.P. cock	On
H.P. cock	OFF

When r.p.m. reach 2,000, release relight button and close throttle fully.

(c) Relighting Derwent Mk. 9 Engines

Altitude	Below 25,000 ft.
Airspeed	Below 230 kts., above 20,000 ft.
Throttle	Closed
L.P. pump	On
L.P. cock	On
H.P. cock	OFF
R.p.m.	Steady

Open H.P. cock and hold in relight button until r.p.m. rise.

(d) If either mark of engine fails to relight within 30 seconds, close H.P. cock and, after 1 minute, attempt to relight at lower altitude.

(e) Do not attempt to relight using normal starter button.

ENGINE FIRE

Throttle	Close immediately
H.P. and L.P. cocks	OFF
L.P. pump	OFF
Airspeed	Minimum

Press extinguisher button. If fire persists abandon aircraft.

JETTISONING SPEEDS

Hood	No restriction, but as low as possible
Ventral tank	345 kts. max.
Wing drop tanks	260 kts. max.

UNDERCARRIAGE EMERGENCY

(a) Hydraulic pump failure
Select DOWN and use handpump.

(b) Selector lever jammed
Throttle back starboard engine, reduce speed and operate airbrakes until selector is free.

(c) Ground-lock in with undercarriage UP
Operate ground lock override, then select DOWN normally.

FLAPS/AIRBRAKE FAILURE
Make normal selection and use handpump.

FORCED LANDING

Glide at	170 kts.
Drop tanks	Jettison
Airbrakes	IN
Undercarriage	DOWN
Fuel	H.P. and L.P. cocks OFF
Flaps	As required
Seat harness	Tight. Release all other attachments
Brakes	Pressure
Hood	Jettison (pilot's discretion)

ABANDONING

Airspeed	Minimum
Throttles	Closed
Seat	Fully lowered
Gunsight	Lowered
Harness	Tight and locked
Helmets	Straps fastened and tight
Hood	Lower head and jettison
R/T lead	Disconnect and stow
Harness	Released
Airbrakes	IN

Dive over inboard trailing edge of wing (port side from front seat unless in spin to the left).

DITCHING

All drop tanks	Jettison
Hood	Jettison
Flaps	Fully down
Undercarriage	Up
Parachute harness	Released

CHECK LISTS

Outside back cover
A.L.4.

CHECKS BEFORE TAKE-OFF

Trim	Elevator: 1 div. nose-up Rudder: Neutral
Airbrakes	IN
Fuel	Contents. L.P. and H.P. cocks fully on. L.P. pumps on. Ventral or wing drop tanks selected. Balance cock closed.
Flaps	Up or ⅓ down. Lever NEUTRAL.
Instruments	Artificial horizon erected. Mk. 4F compass switched on and synchronised. Turn and slip OFF flag not visible, correct functioning. Pressure-head heater ON.
Oxygen	Fully ON. Contents. Emergency OFF. HIGH flow. Connections checked.
Hood	Closed. Pointers in line. Handles ¼ in. from horizontal position. Warning lights out (post-Mod. 1787).
Heating	OFF
Harness	Tight and locked.
Controls	Full and correct movement.

CHECKS BEFORE LANDING

Airbrakes	IN
Undercarriage	Down below 175 kts. Lever fully down.
Fuel	Contents. Balance cock as required.
Flaps	As required.
Harness	Tight and locked.
U/C indicator	Three green lights. Nose-wheel indicator showing.
Brakes	Pressure. On, off, exhausted.

INSTRUMENT APPROACH

DOWNWIND

   11,000 r.p.m., ¼ flap, 150 kts.

BASE LEG:

   11,000 r.p.m., ½ flap. 145 kts.

GLIDE PATH:

   11,000 r.p.m., full flap, 130 kts.

ENGINE LIMITATIONS

Power rating	R.P.M.	J.P.T. C.
Take-off, climb and operational necessity (15 mins.)	14,700 ± 100*	680†
Maximum continuous	14,100	630
Idling on ground (throttle closed)	3,500 ± 200	500

* R.p.m. at sea level automatically restricted to 14,550.
† 700°C. above 20,000 ft. for 10 minutes.

(A9230)

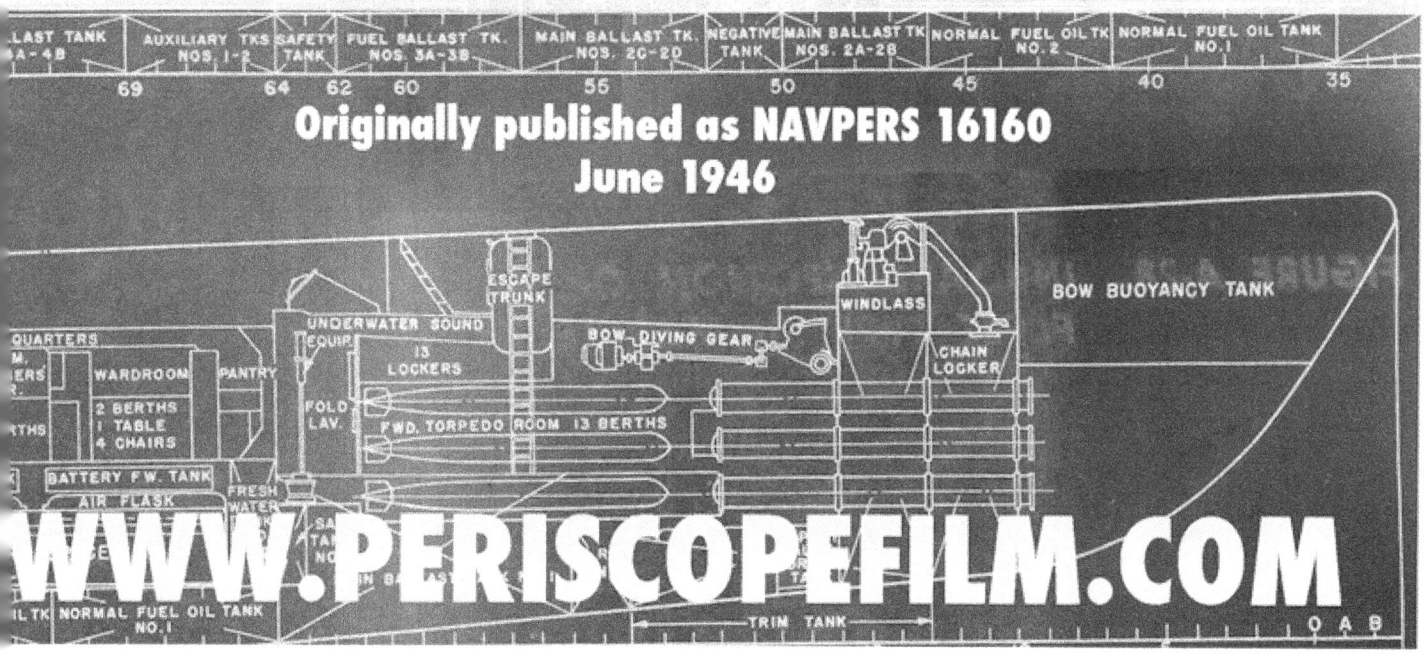

WARSHIPS DVD SERIES

NAVY

AIRCRAFT CARRIER
MISHAPS
SAFETY AND TRAINING FILMS

NAVY

-PERISCOPEFILM.COM-

DVD VIDEO

NOW AVAILABLE ON DVD!

EPIC BATTLES
OF WWII

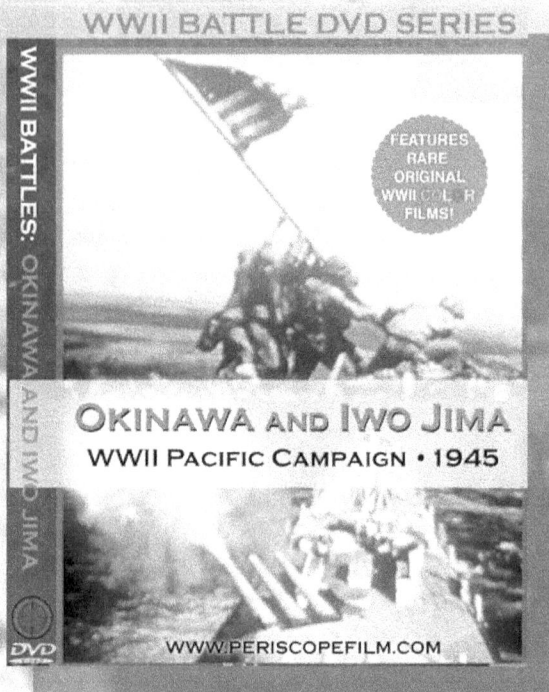

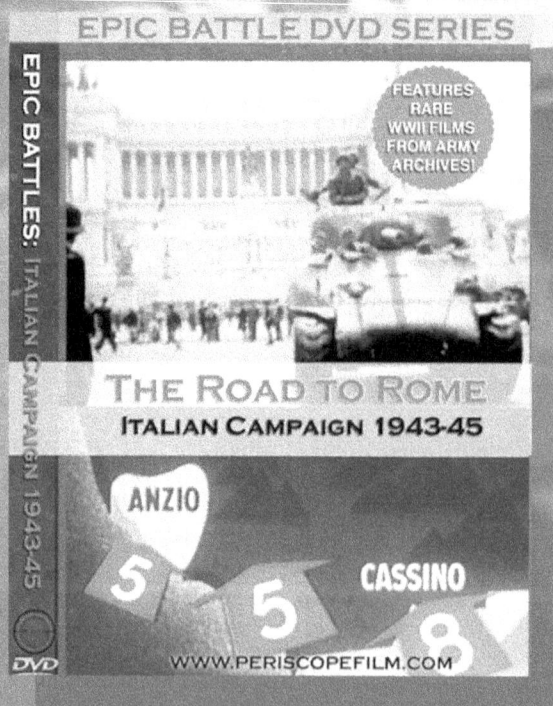

NOW AVAILABLE ON DVD!

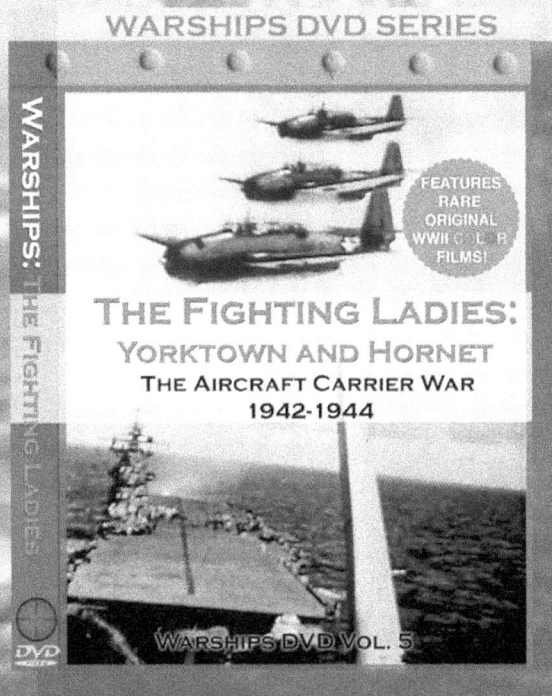

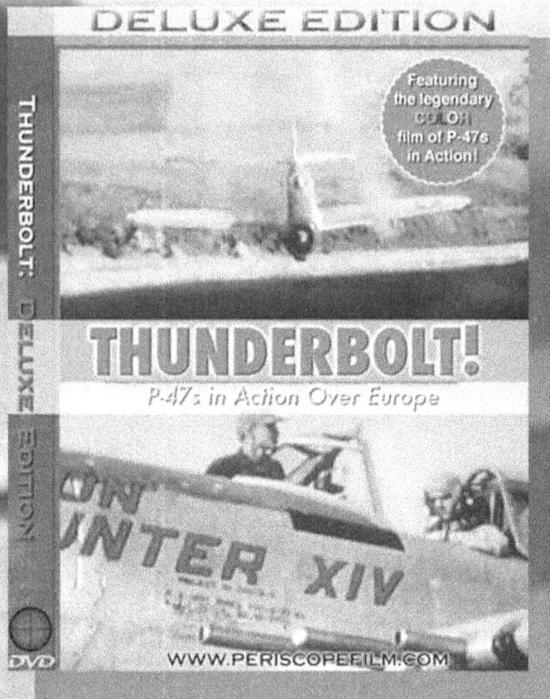

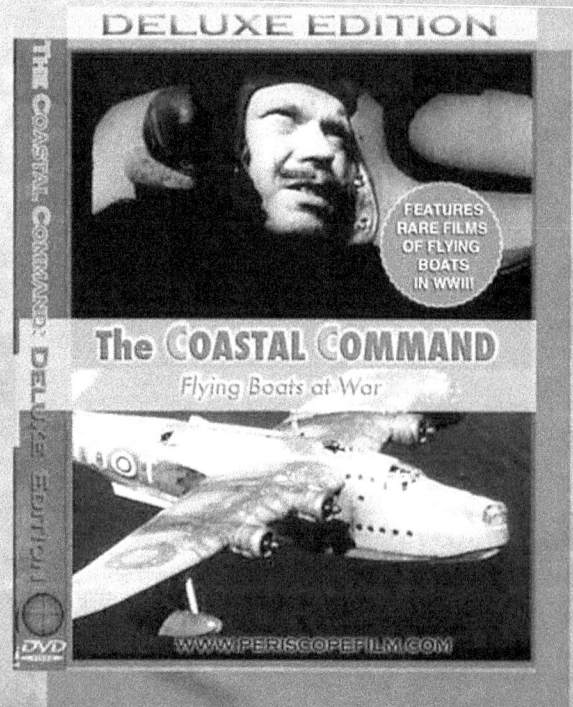

www.ingramcontent.com/pod-product-compliance
Lightning Source LLC
Chambersburg PA
CBHW080519110426
42742CB00017B/3170